SELF-GOVERNANCE IN SCIENCE

Commercial and academic communities use private rules to regulate everything from labor conditions to biological weapons. This self-governance is vital in the twenty-first century, when private science and technology networks cross so many borders that traditional regulation and treaty solutions are often impractical. *Self-Governance in Science* analyzes the history of private regulation, identifies the specific market factors that make private standards stable and enforceable, explains how governments can encourage responsible self-regulation, and asks when private power might be legitimate. Unlike previous books that stress sociology or political science perspectives, Maurer emphasizes the economic roots of private power to deliver a coherent and comprehensive account of recent scholarship. Individual chapters present a detailed history of past self-government initiatives, describe the economics and politics of private power, and extract detailed lessons for law, legitimacy theory, and public policy.

Stephen M. Maurer has taught and conducted research at UC Berkeley's Goldman School of Public Policy since 1999. Trained as a lawyer, he has published more than forty articles in leading journals on topics ranging from innovation economics to national security. He also has extensive practical experience helping academic and commercial scientists organize community-wide initiatives. Maurer is editor and lead author of two previous books, *WMD Terrorism: Science and Policy Choices* (2007) and *On the Shoulders of Giants: Colleagues Remember Suzanne Scotchmer's Contributions to Economics* (2017).

Self-Governance in Science

*Community-Based Strategies for
Managing Dangerous Knowledge*

STEPHEN M. MAURER

Goldman School of Public Policy,
University of California

CAMBRIDGE
UNIVERSITY PRESS

University Printing House, Cambridge CB2 8BS, United Kingdom

One Liberty Plaza, 20th Floor, New York, NY 10006, USA

477 Williamstown Road, Port Melbourne, VIC 3207, Australia

314-321, 3rd Floor, Plot 3, Splendor Forum, Jasola District Centre, New Delhi - 110025, India

79 Anson Road, #06-04/06, Singapore 079906

Cambridge University Press is part of the University of Cambridge.

It furthers the University's mission by disseminating knowledge in the pursuit of education, learning and research at the highest international levels of excellence.

www.cambridge.org
Information on this title: www.cambridge.org/9781316622940
DOI: 10.1017/ 9781316771044

© Stephen M. Maurer 2017

This publication is in copyright. Subject to statutory exception and to the provisions of relevant collective licensing agreements, no reproduction of any part may take place without the written permission of Cambridge University Press.

First published 2017

A catalogue record for this publication is available from the British Library

ISBN 978-1-107-17180-0 Hardback
ISBN 978-1-316-62294-0 Paperback

Cambridge University Press has no responsibility for the persistence or accuracy of URLs for external or third-party internet websites referred to in this publication, and does not guarantee that any content on such websites is, or will remain, accurate or appropriate.

*For the organizers: Leo Szilard, Richard G. H. Cotton,
Markus Fischer, and all the other scientist-entrepreneurs
who proved the skeptics wrong*

Contents

Preface		*page* ix
Acknowledgments		xv
	Introduction	1
	PART I THE FIRST HUNDRED YEARS	
1	Prelude: Self-Governance to 1980	11
	PART II COMMERCIAL SCIENCE	
2	Legacy: The New Self-Governance	23
3	Commercial Self-Governance I: Private Power	48
4	Commercial Self-Governance II: Private Politics	72
	PART III ACADEMIC SCIENCE	
5	Legacy: Academic Self-Governance in Modern Times	99
6	Academic Self-Governance: Power and Politics	119
	PART IV LEGITIMACY, LAW, AND POLICY	
7	Legitimacy	147
8	Law	165
9	Policy and Practice	178
10	Extending the Model	201
	Conclusion	222
Notes		225
References		271
Index		295

Preface

I thought this project would last a year or two.

That was in 1999. I had just written an article on whether Congress should pass new legal protections for science data.[1] Two days after the piece appeared, the late geneticist Dick Cotton sent me an e-mail asking if I would help six hundred mutations biologists negotiate with industry to create a community-wide database.

I felt sheepish, not least because I didn't know all that much about science data. But this seemed like a good way to learn, and the idea of a community-wide initiative intrigued me. Then, too, the project suited me. I had been a business lawyer, so I knew something about transactions. Also, helping Cotton's group would let me try out ideas about how private governance worked. Amazingly for a social scientist, I could experiment.

In the end, I spent hundreds of hours on Cotton's project. As I report in Chapter 5, the results were mixed. The good news is that I quickly saw that success was possible. I suppose that sounds like small beer, but skeptics like to dismiss self-governance as obviously insincere ("public relations") or wrongheaded ("like herding cats"). What I had seen instead was that the great majority of mutations scientists wanted this project to work. If self-governance was a fraud on the public, then these scientists were also victims.

More importantly, I began to see what the "cat-herding" trope missed. When I joined the project, Cotton was still trying to convince colleagues one at a time. At this rate, persuading all six hundred community members really would have been impossible. But as Web pioneer Tim Berners-Lee once said, organizing private standards is like pushing a sled: hard at first, but then progressively easier, until finally all you can do is hang on.[2] I particularly remember one meeting where opponents who had sniped endlessly at Cotton's project began to realize that success was possible – and that they might be left out. Suddenly, in the space of an afternoon, everyone wanted to help. I knew that Silicon Valley standards wars often ended with one side's collapse, but seeing the thing happen in life was amazing.

And yet, despite this, that first experiment deadlocked. In its way, that was even more instructive. Like everyone else in America, I grew up in a society that takes majority rule for granted. But why exactly should that be? The mutations initiative started to unravel when just three out of eighty-some members loudly announced that holding a long-planned vote was "unethical" and would "split the community." On the face of things, the tactic was absurd. Even so, it worked: three hours later, the majority shuffled out of the meeting without ever having called for a show of hands. If I had been an activist I would have been crushed. But I wasn't an activist, I was an observer, and the whole thing intrigued me. The usual democracy rhetoric never says much about dissenters except to praise them. Now I saw that there was more to think about. The sheer weirdness of the outcome hinted at a deeper theory.

Later, I tried again. In 2006, the Carnegie and MacArthur foundations asked my Berkeley project to help a second group of academics – "synthetic biologists"* this time – organize an antiterrorism standard. In many ways, the result turned out to be a replay of the mutations experience, this time with a single dissenter staring down almost two hundred colleagues.

By now I was addicted. In 2007, business executive Markus Fischer asked me to help develop a private standard to make sure that companies that make artificial DNA did not accidentally sell, say, smallpox genes to terrorists. Third time lucky. Partly this was about culture: business executives focus on results, and if the meeting had failed, I think the participants would have cursed the organizers for wasting their time. But other phenomena were more familiar. Once again, there was a dramatic tipping dynamic: once Big Pharma giant AstraZeneca began praising self-regulation, firms that had bitterly opposed private standards frantically reversed themselves. Three months later, Fischer's European trade organization standard and a near-identical American copy had spread across the globe. The thing could work.

So I had my experiments. I saw how they resembled one another. I had puzzles that didn't make sense. Best of all, I could see that the usual bumper sticker reasoning ("public relations," "herding cats") was not nearly good enough. But in that case, I had to wonder what a really correct theory would

---

* Synthetic biology can be adequately if loosely defined as the branch of genetic engineering that uses artificial DNA to perform experiments.

look like. The obvious next step was to fit what I had seen into conventional social science. I began by doing what I should have done all along – chasing the literature in libraries and on the Web.

The more I read, the more I realized how much history really did repeat itself. Academics had already obsessed about "splitting the community" at the 1975 Asilomar Conference on Recombinant DNA, where biologists had taken the first tentative steps to regulate genetic engineering. In the Nineties, big companies had repeatedly bullied suppliers to self-regulate in multiple industries ranging from food to coffee to fisheries. Above all, I saw how often self-governance had succeeded. The food in grocery stores was almost entirely governed by private rules and private timber regulation was roughly as burdensome as the government kind. Most strikingly of all, American physicists had once blocked Nazi Germany's last, best chance at an atomic bomb.

My library phase took a long time. Like many interdisciplinary subjects, the central papers were scattered in strange places – obscure political science journals, specialist law reviews, even forestry periodicals. This made progress much slower than it should have been. Even today, most practitioners still struggle to get their arms around basic facts. As for theory, authors seldom offered more than an offhand sentence or two. Then, too, what they did say almost always drew on sociology and political science. This was not surprising for a subject called "governance." But it ignored a great deal. Surely it mattered that private organizations had replaced government's famous "monopoly of violence" with economic signals. But markets are usually celebrated for giving people what they want. So why did some markets do the opposite by conferring power over others? And how did this shape the way private governance was practiced? Somehow, the subject needed to make contact with economics.

At this point, the reader could be forgiven for thinking that I wrote this book on the Edmund Hillary principle that self-governance is mostly interesting "because it's there." But in truth, there are better reasons. The first is practical. I have already said that most scientists support self-governance. But it is equally true that I have never met a single community member who came to the subject knowing much more than the magic word "Asilomar." This is profoundly disabling: a community that does not know its own history wastes endless time reinventing bad ideas. If anything, the deficit was worse for government officials. Some years after 9/11, the FBI began asking academic and commercial biologists to join "stakeholders conferences," transparently hoping that participants would volunteer to regulate themselves.

But in fact, the agents knew next to nothing about private standards. One FBI agent even told the press that public problems are "not something you can solve in a meeting in a hotel conference room."[3] Anyone who knew even a little history could have told him: for private regulation, that's *always* how it's done.

The second reason to study private governance is intellectual. The founding generation of modern political scientists – figures like Harold Lasswell and Robert Dahl – had rebelled against pre–World War II traditions that equated politics with the lawyerlike study of government and its rules. Politics, they protested, was timeless and universal, something that happened every time humans tried – which sooner or later they always did – to organize collective action. The profession's first efforts to separate politics from institutions had encouraged comparative studies across hundreds of jurisdictions. But those results were more limited than they seem, not least because practically all modern governments either descend or pretend to descend from the same eighteenth-century blueprints. Elinor Ostrom's great contribution in the 1990s had been to expand the universe of case studies to include informal and nontraditional governments. But her examples were almost always preindustrial and depended on norms and cultural facts that had vanished centuries ago. Modern audiences were bound to find these lessons obscure.

The New Self-Governance is different. It exists today. The players look exactly like the rest of us. Best of all, it is inventive: while every large organization claims to be "democratic," it does so through wildly different institutions. This ought to be a godsend for scholars.

My final set of reasons relates to policy. Despite hundreds of empirical papers, scholars still write about the New Self-Governance's law and legitimacy challenges as if the whole subject were hypothetical. This ignores most of what we know about real initiatives. Worse, general problems are hard to solve. Progress might come faster if we focused on the "special cases" that actually exist.

In the meantime, there is a growing urgency. For the past decade or so, it has become conventional for national security scholars to call for private governance.[4] But they have said almost nothing about how to go about organizing such ventures, so that the advice they give to policymakers – that governments should pursue "creative partnerships with stakeholders" and "drive change within the system"[5] – is mostly platitudes. This is unacceptable. Governance scholars have studied this subject for a quarter century now. We can do better.

Looking back, I am glad that I took a practical path into this subject. Aside from the sheer pleasure of the thing, it fed intuitions that I might never have noticed from written sources alone. All the same, it raises obvious dangers of subjectivity. The best answer, I think, is that intuition is only scaffolding. A theory is either right or wrong. It explains the previous literature. It predicts. Or it doesn't. Either way, the final product is what matters.

Despite this, I have also taken two explicit precautions against subjectivity. The first was to remember that a good lawyer focuses on process and solutions but respects his client's goals. This helped me maintain a certain detachment. I think it is fair to say that I never much cared which side won the substantive arguments I witnessed, so long as the debate itself proceeded fairly. The fact that many of the arguments were highly technical made this easier than it sounds.

Finally, I have relied on the usual discipline of scholarship. Practically all of the facts that follow are footnoted in the conventional way. Granted that I have occasionally added my own testimony; this is mostly in the spirit of entertainment. Nothing depends on these additional facts, and the careful reader will find that she arrives at the same arguments just as quickly without them.

I thought this project would take a year or two. That was eighteen years ago. Now the book is in your hands. Here's hoping you enjoy the trip.

Acknowledgments

Many friends and colleagues helped make this book possible. My special thanks to biologists Markus Fischer, Jay Keasling, and the late Dick Cotton, who invited me to participate in their respective communities; Jean Tirole, Jacques Cremer, and Paul Seabright, who hosted me for a memorable spring 2013 semester thinking about these issues at the Institute for Advanced Studies in Toulouse; the MacArthur and Carnegie foundations, which provided financial support for my work with academic and commercial synthetic biologists; John and Linda Gage, who supported my program at UC Berkeley, including, crucially, my earliest work with mutations scientists; and UC Berkeley's Center for Information Technology Research in the Interest of Society (CITRIS), which generously provided office space to complete the project.

All worthwhile scholarship is partly communal. I especially thank Tim Lytton for his many insights and, more materially, sharing his extensive collection of governance articles and hosting the only study group that unites an otherwise scattered community; Errol Meidinger, the subject's deepest theorist, for the idea of external competition; Sebastian von Engelhardt, who brought an economist's tools to modeling how private power emerges from commercial supply chains; Gretchen Hund and Andrew Kurzrok for their deep knowledge of nuclear nonproliferation policy; and above all, the late Suzanne Scotchmer, who encouraged me to pursue self-governance research and later provided the characteristically crucial insight that even cartelized industries find it hard to escape quality competition. I also wish to thank Eugene Bardach, Jens Carsten, Andreas Freytag, Rob Van Houweling, Revati Kapshikar, Stefan May, Robert Mikulak, Florian Möslein, Michael Nacht, Karl Riesenhuber, Craig Shank, Eric Schickler, Tom Slezak, and Ray Zilinskas for their many comments and suggestions.

And of course, to my unfailingly cheerful Cambridge University Press editor, Karen Maloney, along with the anonymous referees who greatly improved the manuscript.

With this kind of help the current book ought to be perfect. It is not. The errors and omissions are mine alone.

Introduction

The New Governance is twenty-five years old. Anyone who follows the various notes in this book will see just how much has been written. There are now hundreds of case studies and empirical papers, and something like a consensus about what they say. But the careful reader will also see the literature's limits: in the words of one recent scholar, the insights so far mostly come from "stitch[ing] together and scal[ing] up" best practices from case studies.[1] Worse, this semiempirical approach "seem[s] to have hit a ceiling."[2]

To be sure, practically all academic fields reach such moments. But it does suggest that it is time for scholars to take stock. To look for some explicit, first-principles theory that accounts for what we have learned. And, above all, to try to see the subject whole.

Self-Governance in Science

We begin by defining our topic. Older methods for organizing "self-governance" traditionally relied on the fact or threat of state intervention ("shadow of hierarchy"), industry efforts to fix market imperfections, or the need to maintain "interoperability" among products. We mostly address these in Chapter 1. Like most recent authors, most of what follows will instead focus on the "New Self-Governance", which first became prominent in the early 1990s. It typically exploits Western firms' power over global supply chains to impose elaborate and sometimes burdensome regulation on large numbers of people. By now, scholars agree that this variant differs from earlier models along at least four dimensions:

Independence. Private initiatives exercise power without any support from traditional governments[3] beyond basic law and order.[4]

Coercive power. For any sufficiently large group, there will always be at least one dissenter who prefers not to cooperate. At this point, collective action becomes impossible unless and until the majority can compel his participation. To this extent, "private governance" is inseparable from coercion.[5] Conventional distinctions between "legitimate" and "illegitimate" power, or "the consent of the governed" mostly evade the deep legitimacy issues that arise when private bodies make "binding decisions ... that affect the quality of life and opportunities of a wider public" that seldom show much interest or ability to participate.*

Permanence. Actors seek to create enduring organizations in which adaptation, inclusion, and learning can occur over time. The new organizations invariably embrace a wide range of stakeholders.[6]

Impact and scope. Scholars broadly agree that private regulations effect "real change" on the ground.[7] Compared to earlier examples, the New Self-Governance typically seeks out problems that are only distantly related to earning a profit. Examples include environmental protection, national security, working conditions, and other topics formerly dominated by traditional governments.[8]

Finally, one might expect a book on "self-governance in science" to limit itself to communities whose memberships are dominated by working scientists. In fact, we usually cast a wider net, including any private group that seeks to engage issues that are "scientific" in the sense that Environmental Protection Agency (EPA) regulations say, depend on science. In practice, science is nevertheless always central to our examples, if only because, in the words of one prominent businessman, technology companies "are the ones in constant danger of committing violations."[9] In any case, excluding these examples would discard too much evidence about how private governance operates in the modern world.

The Basic Argument

Many social goals are additive; given enough small, individual contributions, success can be achieved. But others are multiplicative: if even a

---

* Rudder (2008) at pp. 901, 913. The fact that self-governance bodies set rules for others takes them beyond the much simpler "contract governance" case where members suspend or rewrite public law for themselves alone. Since no outsiders are affected, contract governance unambiguously improves welfare for the entire society. Calliess and Renner (2009) at p. 1343 (commercial law "may just as well be performed by private governance regimes" but that situation becomes "considerably more complicated" where transactions affect the public interest).

few members refuse to cooperate, the value of everyone else's compliance declines steeply. At this point, collective action becomes the only option.

But collective action brings its own problems. For any large body, the chances of unanimity are negligible. This forces organizers to overrule dissenters. This coercion – there is really no other word for it – comes in many varieties. The most familiar is jail and physical force – Max Weber's famous "monopoly of violence."[10] The difference for private bodies is that the coercion never escalates beyond financial incentives. All the same, this is often sufficient. The distinction is especially thin for entities, where prolonged negative profits really do inflict a kind of death penalty.

We can guess one centrally important fact immediately. Since collective action is the opposite of individual choice, private power cannot be reconciled with the textbook ideal of perfectly competitive markets. Instead, this book argues that real (imperfect) markets often give some private actors power over others; that this power is sometimes sufficient to enforce industry-wide rules on suppliers; and that private politics can be understood as an extended struggle to create coalitions big enough to impose such rules. It then generalizes this baseline commercial model to academic communities, building on standard economic theories in which contracts and simultaneous exchange are replaced by trust relations (e.g., peer review) in which parties trade reciprocal favors over time.

A second set of arguments adds that, for some industries, oligopoly competition forces firms to implement consumers' policy preferences in much the same way that elected officials worry about public opinion. This "shadow electorate" dynamic has important implications for democratic legitimacy, legal doctrine, and public policy.

The crucial decision, of course, is what theoretical framework to use. This book adopts the simplest possible choice, deploying classical microeconomics and two-person game theory to ask when and how anchor firms can impose industry-wide rules on suppliers.* Though phrased in economic

---

* The existing literature has so far done remarkably little to develop this economics. Probably the most extended examples are due to Prof. Gereffi who characterizes New Governance models as "buyer-driven" systems, but leaves the microeconomics behind this observation off-stage except for offhand remarks, e.g., that big firms possess "buying power" (Gereffi 1994 at p. 115) or that global supply chains can be usefully analyzed as a "production system" that "links" firms' "economic activities" to "technological and organizational networks." (Gereffi, Humphrey, and Sturgeon 2005; Gereffi 1999; Gereffi 1994); cf. Locke (2013) at p. 28 (starting from the premise that "if all powerful global brands are willing and able to dictate commercial terms ... on their 'weak' and/or 'dependent' suppliers, they must also be capable of forcing these same suppliers to comply with codes of conduct ..." (id. at p. 28).

language, this latter threshold is also implicitly political, since knowing how many anchor firms must agree to new rules is already a kind of constitution.

But of course, politics is more than constitutions. We also require some theory of how actors try to influence the community to adopt some standards and not others. Once again, the book's approach is deliberately conventional. We start with the simple coalition models first introduced by Anthony Downs.* These nicely explain which coalitions would emerge if initiative members had perfect information. At the same time, real private politics outcomes are chaotic and contingent. For this reason, Downs's determinate models cannot be a complete description. We therefore extend the analysis to include Oliver Williamson's insights into organizations whose decision-makers suffer from asymmetric and "impacted" information.[†] Like most theories that emphasize friction, our analysis does little to predict the outcomes in specific cases. It can, however, offer useful insights into why private governance produces such random results, and how better institutions could improve the situation.

Plan of This Book

This book reviews the history of private governance, uses this evidence to construct natural and parsimonious theories of the phenomenon, and then asks what these models have to say about legitimacy, law, and public policy. The balance of this section provides a detailed road map of what readers can expect.

Part I (**The First Hundred Years**) introduces the history of self-governance from the rise of nationwide institutions in the late nineteenth century through the 1980s.

> Chapter 1 **Prelude: Self-Governance to 1980.** This chapter recounts the "old" self-governance from the 1890s to 1980 or so, including examples from commercial and academic communities. On the commercial side, this includes "shadow of hierarchy" bodies that traditional governments encouraged for their own ends; private bodies formed to make markets feasible and less costly; and projects designed to create shared industry assets. We then turn to private academic governance, focusing on American physicists' wildly successful conspiracy to deny nuclear fission data to Nazi Germany.

---

* Downs (1957).
† Williamson (1975).

Part II (**Commercial Science**) completes the historical account of commercial governance by describing the New Self-Governance that grew up around globalization in the 1990s. It then generalizes from these examples to develop a theory of private politics in commercial markets.

> Chapter 2 **Legacy: The New Self-Governance.** This chapter builds readers' intuition by recounting the history of the New Self-Governance in seven industries.* Theory-inclined readers may prefer to read Chapter 3 first before diving into this material.
>
> Chapter 3 **Commercial Self-Governance I: Private Power.** This chapter collects and stylizes the accounts in Chapter 2 to present an explicit theory of when and how markets create private power. Here the institutional actors include *both* the anchor firms that dominate global supply chains *and* the various suppliers, NGOs, and other actors to whom this power is sometimes delegated. The chapter ends by arguing that anchor firms often adopt private standards to defend market share. This can sometimes elevate consumers into a "shadow electorate" whose preferences constrain private discretion in much the same way that real electorates limit elected officials.
>
> Chapter 4 **Commercial Self-Governance II: Private Politics.** The last chapter of Part III develops a theory of how private politics decides when and if private power is exercised. The discussion starts from Downs's classic theory of coalition politics before expanding to include Williamson's "information impactedness" case, where members must decide based on incomplete information. The chapter closes by exploring how attempts to persuade the shadow electorate resemble and sometimes depart from conventional mass politics.

Part III (**Academic Science**) generalizes and extends Part II's commercial analysis to include academic self-governance. It describes four representative examples of academic self-governance and uses this foundation to construct an economic theory of private academic power in analogy with Part II's analysis of commercial markets. It then explores how private politics emerges to deploy this power in modern academic science communities.

> Chapter 5 **Legacy: Academic Self-Governance in the Modern Era.** This section begins by describing the Asilomar conference (1975) and three more recent attempts at academic self-regulation. Theory-inclined readers may prefer to start with Chapter 6 and return to this history later.

---

* Tuna, fisheries, food, lumber, coffee, nanotechnology, and artificial DNA.

Chapter 6 **Academic Self-Governance: Power and Politics.** Academic scientists routinely allocate resources to each other using peer review and other, similar methods. We argue that these nonmarket trades are best understood through the large economic literature on trust and reputation games. In particular, dissenters' recurring threat that collective action will "split the community" points to a real possibility that established patterns for allocating resources will be disrupted or become less predictable. This leads to a characteristic politics in which small groups of players can sometimes block the majority.

Part IV (**Legitimacy, Law, and Policy**) asks whether and to what extent private self-governance is legitimate. This sets the stage for the book's law and policy discussions.

Chapter 7 **Legitimacy.** This chapter summarizes the large literature on when private self-governance organizations should be considered legitimate in the absence of conventional elections. It argues that executives in oligopoly markets often feel strong pressure to please consumers, so that the latter become shadow electors in shaping private policy. When this happens, legitimacy depends on whether institutions are sufficiently transparent to make this control effective. We also consider alternative models in which democracy is enforced by traditional government ("shadow of hierarchy"), debate ("deliberative democracy"), or widespread public support ("outcome legitimacy").

Chapter 8 **Law.** The Sherman Act (1890) and its European and Japanese descendants were originally conceived to limit private power. This chapter documents how the interpretation continues to shape modern competitions law and use legitimacy analysis to suggest reforms that could clarify and improve existing doctrine. It then turns to more recent cases that identify competitions policy with economic efficiency, offering practical guidance on how private initiatives can be structured to avoid objections.

Chapter 9 **Policy and Practice.** This chapter asks when and how officials should intervene to promote private governance in commercial and academic settings. It concludes by suggesting a practical agenda for making scholars' work more directly useful to policymakers.

Chapter 10: **Extending the Model.** This chapter asks where self-governance could spread in the future. In contrast to earlier chapters' mostly theoretical focus, it focuses on the nuts-and-bolts details of how private governance could be established in particular industries. It argues that many of the most promising opportunities relate to "dual-use" civilian technologies that can also be used to make weapons. The chapter ends by describing how academic scientists can use self-governance methods to manage emerging technologies and build shared infrastructure.

Conclusion. The book concludes by suggesting some priorities for future research.

PART I

THE FIRST HUNDRED YEARS

1

Prelude: Self-Governance to 1980

Like most good magic tricks, self-governance can be done several ways. Three of the most important methods were pioneered at the end of the nineteenth century.

1.1. Industry Precedents

The use of private institutions to pursue public policy is as old as commerce. Indeed, medieval guilds already claimed to protect society's interests. Governments were quick to see that this offered new possibilities for control, most obviously when medieval French kings saw that it was easier to let the University of Paris decide which books ought to be suppressed. More than that, the university used its buying power to drive uncooperative publishers out of business. Remarkably for those days, state violence was never invoked.[1]

1.1.1. Traditional Governance Models: 1890–1980

The rise of Big Business in the nineteenth century expanded these beginnings, showing how private power could be rooted in markets. The earliest examples almost always addressed flaws that made markets unnecessarily costly or impractical.[2]

Fixing Market Defects. In theory, the new nationwide capital markets let factory owners spread risk. At first, however, insurers had no way of knowing what, if anything, their insureds were doing to prevent fires in the first place. The insurance industry fixed this otherwise fatal flaw by drafting detailed fire codes and hiring private inspectors to see that they were followed.[3] This eventually led to the rise of a new private body (Underwriters

Laboratories, or UL) as an independent, self-supporting audit service.[4] While participation was formally optional, companies that failed to use UL regularly lost market share.[5]

UL, in turn, realized that it could deliver more value – and command higher fees – by making sure that its actions were transparent. This led to elaborate procedures modeled on traditional democratic governments, including oversight councils to review test results, explaining decisions in detailed written reports, launching internal investigations in response to criticism, and opening its facilities so that the public could watch tests being performed.[6] Modern scholars give UL credit for creating "a new, professionalized safety industry" far ahead of government, which belatedly codified UL's private standards decades afterward.[7] However, economic incentives were even more important. Crucially, UL had to balance the needs of manufacturers (who paid its fees) and fire insurers (who required UL certification in the first place).[8] This encouraged UL to be creative in meeting manufacturers' needs[9] while still refusing compromises that would alienate insurers.[10]

Most early industry schemes addressed similar market flaws. These included certification schemes that helped consumers judge the quality of goods, services,[11] and promises.[12] These standards were simple to enforce since the industry – including potential violators – found it nearly impossible to survive otherwise. That said, the regulations were tightly focused on markets and largely ignored social issues.

Shadow of Hierarchy. Other industries returned to older "shadow of hierarchy" models in which government gave private standards the force of law[13] or else threatened regulation if private industry failed to govern itself.[14] Standards entrepreneurs often exaggerated these threats for their own purposes, warning that government would intervene if industry did not[15] and dramatizing the risk of lawsuits.[16]

Supply Chain Governance. A very different solution took advantage of large firms' power over their supply chains. The most important early example was the Fashion Originator's Guild of America (FOGA),[17] whose members produced 60 percent of all quality women's wear sold in the United States.[18] In depths of the Great Depression, FOGA announced that its members would no longer sell to "unethical" retailers.[19] Since most retailers needed these goods to remain competitive,[20] at least 12,000 retailers "cooperated" in the scheme – more than half protesting that they had been coerced.[21] Soon, FOGA was running its own intellectual property system,[22] conducting random compliance audits of retailers,[23] and holding trials and appeals for violators.[24] This was too much for the U.S. Supreme Court, which struck down the arrangement on the ground that American antitrust law banned such private "tribunals for [the] determination and punishment of violations."[25]

But private standards were too useful to go away. Lawyers and judges soon found that the FOGA decision's vague rule against "tribunals" left plenty of room for standard setting. By the late 1980s, the United States had an estimated 25,000–50,000 private standards employing roughly 100,000 people.[26] This led to a lively debate on how decisions should be made. Probably the most insistent concept was "consensus," sketchily defined as "much more than a simple majority, but not necessarily unanimity."* Other rules implemented a "general rubric of 'due process'[27] and appeal rights."[28] These included rules favoring broad participation and the use of proxies for those unavoidably absent,[29] rules for appealing and overruling "no" votes[30]; open meetings and membership; notice and comment procedures[31]; and written decisions[32] and records.[33]

As Andrew Russell[34] has emphasized, most of this structure originated with standards evangelist Paul Gough Agnew in the twenties. However, standards bodies seldom if ever explained in any explicit way how their various procedures were supposed to advance democratic ideals or even fairness. Instead, they were allowed to proliferate, in Robert Dixon's phrase, like Amish "hex signs"[35] – i.e., poorly understood features that could not hurt and might possibly improve standards making.[36] Current scholars concede the point, but argue that these "pragmatic steps" are nevertheless "sensible approaches to developing democracy in an arena where the very meaning of the concept is in doubt"[37] and could evolve into new and better institutions over time.[38]

1.1.2. Traditional Theories

We end this section by reviewing how scholars have traditionally made sense of private governance. The first and most common interpretation, as we have seen, was to posit a *shadow of hierarchy* dynamic in which governments ordered private communities to self-govern and threatened

---

* Hamilton (1983) at p. 463. The dominant American Society for Testing and Manufacturing defined "consensus" as two-thirds of the combined negative and affirmative votes at the subcommittee level with at least 60 percent of the voting interests participating, and nine-tenths of the combined negative and affirmative votes at the committee and society levels, with at least 60 percent participation at the committee level and at least fifty votes participating at the society level. Id.

to develop its own rules if they refused. Because government willed private power into being, the resulting arrangement was as legitimate as any other official action. Even so, this left little room for private actors' wishes. Any government that cared enough to intervene in private governance almost always had strong feelings about what ought to be done. To the extent that self-governance did occur, it mostly entered through regulators' inattention.[39]

A second, more subtle model depended on *network markets*, where consumers strongly preferred that products work together, i.e., be "interoperable."[40] But in that case everyone *also* knew that only one standard could survive in the long run. This led to "tipping" dynamics in which consumers would abruptly rush to whichever standard was expected to win. At this point, firms that backed losing standards faced a harsh choice: join the dominant standard or leave the market.[41] Crucially, network markets are agnostic: in principle, many different standards are stable.* Communities often made their choice through standards bodies and private politics. The telecommunications and electrical-equipment industries have picked winners this way for over a century.[42]

Network industry models experienced an unexpected renaissance in the 1990s, when the private World Wide Web Consortium (W3C) was lionized for its "philosopher-king" model of setting Web protocols.† The basic idea was that tipping gave leaders significant though not overwhelming power to decide which standards would win. Early in the race, the king had to persuade followers that his preferred standard was worthwhile. But once tipping set in, the king could safely announce a winner, knowing that the standard could now "go ahead despite objections of a minority."[43] Thereafter, any remaining dissenters would have to drop their objections or

---

* This is a fundamental departure from traditional microeconomic models in which the market makes the single best choice from those available.
† Journalist Mark Fischetti provides a thoughtful account of how the philosopher-king system is supposed to work:

> [T]he few industry people who are critical of W3C claim that Berners-Lee is a king who holds an iron hand over his puppet regime. But his subjects disagree. ... "The question is," Berners-Lee acknowledges "... could I by whim pervert the course of justice? No, because there would be an outcry. I have to put my ideas into the process like anyone else. ... If one succeeds, fine, otherwise members will tell me it's stupid." Most often, the community embraces his ideas. ... Perhaps Berners-Lee plays more of a King Arthur role, sitting at the Round Table with the best technicians who also hold the right social ideals. Fischetti (2009).

go out of business. But this was only the beginning: Given that most community members had limited information, kings who made good choices were more likely to be trusted and gain influence in the future. This influence was even stronger to the extent that some members started to follow the king's endorsements on faith. But if you were a king yourself, actually exploiting this power required very difficult judgments of just when you could declare a consensus and make it stick.

The question remains whether the philosopher-king model is efficient or democratic. Given that standards wars imply a certain amount of randomness, it seems reasonable that a king can improve outcomes on average. But that presupposes a wise and a benign ruler. There is no particular reason to expect this beyond the weak constraint that kings who consistently displease their communities will eventually lose followers.

Enforcement. So far we have concentrated on asking when firms promise to obey common rules. But most critics have a different objection: the problem, they insist, is that firms cannot be trusted to keep their promises in the first place.[44] Here the definitive analysis is due to Carl Shapiro, who explores a model in which firms that practice high standards earn a premium from consumers. But since compliance is expensive, firms that only pretend to follow the standards can earn even more profit until they are caught. Shapiro shows that enforcement is nevertheless effective in two cases. First and most obviously, consumers might watch so closely that cheating is immediately discovered and earns nothing. Second, the premium could be so large that its present discounted value exceeds the one-time profit from cheating.* Carlo Scarpa has shown that a similar analysis also applies when enforcement is delegated to third-party auditors, who have a similar temptation to earn a one-time profit by quietly defunding effort.[45]

This analysis confirms the common intuition that many self-governance schemes are vulnerable to cheating. But it is equally true that the Shapiro and Scarpa models both assume that enforcers receive no direct benefit from compliance. Conversely, cheating makes far less sense when enforcers

---

* Shapiro (1983). The effect is mitigated when firms that honor the standard are willing to report rivals who cheat. Gunningham and Rees (1997). The counterargument is that companies can potentially earn still higher rewards by reciprocally ignoring each others' noncompliance. Enforcement incentives are decidedly stronger for third parties that directly profit from adherence, for example where insurance companies try to suppress "bad risks." *Id.* This explains the usual rule of thumb that firms closest to the consumer – and therefore most vulnerable to backlash – should pay for audits. Roberts (2012) at p. A-233.

expect their own businesses to suffer if the standard is violated. We argue in Chapter 4 that the New Self-Governance meets this criterion.

1.2. Academic Self-Governance

Advocates of academic self-governance almost always start from the example of Asilomar, where molecular biologists asked the federal government to regulate their work.* This tends to overshadow an earlier and much more spectacular success in which physicists helped stop Nazi Germany from acquiring an atomic bomb. The example is particularly illuminating because it dates from an era when the U.S. government mostly ignored academic science so that the "Republic of Science" was left to govern itself.

1.2.1. The Atomic Bomb Conspiracy (1939–1940)

Hungarian physicist Leo Szilard realized that a nuclear chain reaction could lead to atomic bombs as early as 1933.[46] The prospect was ominous in a world where the Nazis had just taken power. Two years later, he began agitating for an agreement to limit experimental data to England, America, and a few other countries. But his colleagues resisted, objecting that Szilard's physics arguments were unworkable (which was true at the time), that scientific secrecy was abhorrent, and that censorship would impede research. Some added that Szilard, who had taken out a patent, was tainted by commercialism.[47]

The Munich Crisis and German scientists' discovery of fission made Szilard's case more pressing from 1938 onward. But when he approached colleagues at Columbia University, that year's Nobel laureate Enrico Fermi expressed strong skepticism that chain reactions would work.[48] Despite this, Szilard nevertheless wrote to another recent Nobelist, Frederic Joliot in Paris, saying that Columbia was worried that atomic bombs might be possible and Fermi was investigating. If Columbia decided to limit publication, Joliot should do the same.[49] This tentative feeler collapsed a few weeks later, when Fermi himself wrote to Joliot saying his group was trying to understand uranium fission and, he assumed, so was everyone else. This was false: we now know that only the Columbia and Paris labs were competing. The immediate

---

* The reverence that scientists and journalists reserve for Asilomar is surprising: if the trucking industry had demanded similar regulation, journalists would surely have accused it of trying to "capture" regulators. In any case, petitioning the government seems like a limited kind of self-governance. From this standpoint, Asilomar is mostly interesting for the community's successful efforts to suppress certain experiments until the government could act. We tell this story in Chapter 5.

implication, however, was that Fermi felt free to publish. By March 1939, the Columbia and Paris groups had both submitted manuscripts confirming that uranium fission did indeed produce neutrons.[50]

Then Hitler annexed Czechoslovakia. Szilard and physicist Edward Teller again urged Fermi to reconsider. Fermi considered secrecy repellant, but refused to pull rank: "After all," he said, "this is a democracy." If the majority was against publication he would go along. He then asked *Physical Review* to withhold publication, only to learn that Joliot had already published in *Nature*. Fermi argued that there was now no secret to keep, but Szilard pointed out that (unlike Joliot) the Columbia paper contained crucial information about the number of neutrons released. Fermi was unconvinced, but put the matter to his group's administrative head, George Pegram. Pegram delayed.[51]

Szilard next persuaded Joliot's collaborator, Victor Weisskopf, to cable Paris saying that the Columbia group would delay publication on the number of neutrons if Joliot did the same. Everyone could go on submitting papers as usual but delay printing; in the meantime manuscripts would continue to circulate among cooperating laboratories in the United States, England, France, and Denmark.* Szilard also persuaded English experimental physicist Patrick Blackett to lobby *Nature* and the Royal Society's *Proceedings* to join the scheme. Finally, Szilard, Teller, Weisskopf, and Eugene Wigner approached senior Danish physicist Niels Bohr. Like Fermi, Bohr doubted that a bomb was possible and argued that it would be hard to suppress truly important results in any case. Nevertheless, Bohr warned his home institute to check with him before publishing.[52] Szilard and his colleagues also spread the proposal to Merle Tuve (Carnegie Institution for Science), Maurice Goldhaber (University of Illinois), E. O. Lawrence (University of California, Berkeley), and the editor of *Physical Review*. This last was crucial: at the time, nearly all nuclear physics papers passed through *Physical Review*'s offices.[53]

Joliot thought that atomic bombs were a distant prospect, disliked secrecy, and thought that any experimental results would either leak or be independently discovered by German physicists in any case. But above all, he worried that others would publish if he did not.

It didn't help that Szilard and Teller were relatively obscure. Despite this, Joliot discussed the matter with colleagues. He wrote back on April 5 citing a rumor that Tuve's Carnegie group had already achieved similar results, apologetically adding that the Szilard/Teller proposal was "very

---

* Szilard also proposed creating a special fund to increase young scientists' salaries as compensation for lost publication opportunities. Weart (1976) at p. 29.

reasonable but comes too late." Szilard cabled back on April 7 explaining that the Carnegie rumor was false and adding that Tuve had joined the embargo. Dropping his earlier excuse, Joliot published anyhow on April 22. Columbia then followed suit. We now know that this persuaded the British and German governments to launch secret atomic energy programs.[54]

French and British scientists stopped publishing fission papers when Germany declared war that September. But when physicists asked the U.S. War Department to impose censorship, it refused, declaring that scientists would have to do the job themselves. Szilard held back his own papers and persuaded one of Fermi's graduate students to do the same.[55] The turning point came when Szilard asked Fermi to suppress a new experiment on carbon cross sections later that spring: "Fermi really lost his temper; he really thought that this was absurd." But Columbia's administrative director finally asked Fermi to keep the work secret.

Szilard also persuaded Louis Turner (Princeton) to delay a paper describing the plutonium path to nuclear energy. When physicists Philip Abelson and Edwin McMillan (UC Berkeley) published some of this same information, they drew howls of protest from scientists as far away as Britain.[56]

Now that scientists had begun suppressing papers, they needed an institution to review new work. In June, Gregory Breit (Wisconsin) persuaded his National Academy of Sciences (NAS) colleagues to organize a censorship body. NAS reluctantly agreed, making Breit chair of its uranium subcommittee. By then, France had fallen. Breit immediately wrote to journal editors asking that all papers first be submitted to his committee; sensitive papers would be restricted to a limited number of workers, with formal publication embargoed until the end of the war. The editors agreed, albeit with "raised eyebrows." Working with Fermi, Harold Urey, Wigner, and others, Breit finally established "total censorship" of American fission research.[57]

Looking back, the embargo came just in time. Lacking the American results, German scientists ended up pursuing a far more difficult, heavy-water path to plutonium. This decision goes a long way toward explaining why Nazi Germany never built a working reactor.

1.2.2. First Impressions

American academics' conspiracy to suppress atomic bomb data took place in an era when Big Science was mostly funded by individual universities and medical charities.[58] The modern federally funded system was still five years away. Despite this, the physicists' conspiracy teaches some important lessons. The first is that private self-governance is possible, even when government itself is indifferent. At the same time, success was a near thing. Misunderstandings and small accidents of timing could easily have derailed the effort, particularly given the secrecy and competition between labs. The initiative was also lucky to have had someone as skilled and passionate as Szilard for its leader.

Second, community was central. Even prominent players like Fermi refused to take unilateral action. This gave junior people like Teller and Szilard the leverage they needed to raise the issue. At the same time, self-interest ran deep. However necessary, censorship posed a direct threat to individual groups' need to publish and members' career prospects. The fact that restrictions were temporary rather than permanent reduced but did not eliminate this tension.

Finally, Szilard and Breit showed real genius by enlisting *Physical Review* and *Nature* in their conspiracy. This was only partly about having the physical ability to block publication. Unlike bench scientists, journals were semi-independent bodies that stood outside the community. This made it easier for editors to take stands even when some members objected.

PART II

COMMERCIAL SCIENCE

2

Legacy: The New Self-Governance

This chapter continues our history of self-governance, emphasizing the New Self-Governance methods pioneered by commercial communities in the 1990s. To some extent, this focus is conventional and mirrors the academic literature. However, there is also a didactic purpose. We will see that the economic roots of private power are unusually simple in the case where large retailers and manufacturers (collectively, "anchor firms")[1] manage elaborate global supply chains. We therefore start with this baseline case, deferring the more complicated economics of academic communities to Part III.

2.1. Dolphin-Safe Tuna

The rise of globalization in the 1980s established new supply chains that coupled the buying power of large Western retailers with sophisticated worldwide compliance networks. While the system was originally constructed to extract better price and quality terms, the same power could be used to impose almost any behavior on suppliers. The past twenty years have exploited this power to organize increasingly ambitious rules that are only distantly related to profitability.[2] These include hundreds of standards regulating everything from working conditions to energy efficiency to "conflict diamonds."[3] Most of these standards are limited to individual retailers. For large buyers like Walmart, this already affords government-like power. However, it is natural to ask when collective action makes sense, i.e., when it is both feasible and useful to extend individual anchor-firm rules to entire industries.[4] This has become the defining feature of the New Self-Governance.[5]

Saving Dolphins. Newspapers began carrying accounts of tuna fishermen killing dolphins in the 1980s.[6] Fearing that consumer backlash would depress sales, StarKist – the world's largest processor – reached out to nongovernmental organization (NGO) Earth Island Institute to develop a dolphin-safe standard in 1990.[7] The company then announced that it would only purchase tuna caught according to its protocol. Competing canners Chicken of the Sea and Bumble Bee adopted identical policies almost immediately. Together, the three companies accounted for 84 percent of U.S. canned tuna sales and nearly half the world market.[8]

In principle, suppliers could have built one set of facilities for dolphin-friendly customers and a second set for everyone else. In that case, controls would have covered perhaps half the fishing fleet. In practice, it was cheaper to build a single distribution channel for all purposes. By 2011, 471 companies in 67 countries had joined StarKist's standard. These included The U.S. Tuna Foundation, commercial fishermen, small canneries, brokers, import associations, retail stores, and restaurant chains[9] comprising about 90 percent of the world canned-tuna market.[10]

The new rules forced costly changes to industry practice. Compliance was enforced by a kind of miniature government in which suppliers paid Earth Island to inspect canneries, docks, and fishing vessels.[11] By 2007, worldwide dolphin mortality from fishing had been slashed from more than 80,000 to just 3,000 per year.[12] Meanwhile dolphin-safe labels helped increase canned tuna's market share.[13]

Government interactions with the private standard were complex. Some foreign suppliers probably joined hoping to preempt official U.S. trade embargoes.[14] Despite this, shadow-of-hierarchy effects seemed to be weak. For one thing, the Earth Island standard continued to operate even after Congress passed its own, sometimes conflicting standard later in the decade.[15] For another, worldwide fishing practices barely changed when the World Trade Organization struck down the U.S. restrictions in 2012.[16]

2.2. Food Safety

Many large European supermarkets responded to the "mad cow" scandal by imposing detailed handling and inspection requirements on their suppliers.[17] This encountered little or no resistance. The reason, almost certainly, was the economic power of anchor firms in an industry where suppliers typically spend years building relationships with a "huge retailer" that they "can't afford to lose."[18] At the same time, the proliferation of standards forced suppliers to maintain redundant facilities[19] and limited the number of suppliers that could bid for orders.[20] This created an obvious financial incentive to develop unified rules. Starting in 1996, the Association of UK

Retailers began work on an industry-wide food code[21] that was adopted throughout the UK[22] and Holland.[23] Similar retail food association initiatives led to the German/French International Food Standard in 2002–03[24] and the American Safe Quality Food standard in 2003.[25]

These regional standards harmonized practices at the national level but remained a drag on long-distance trade. In 1997, thirteen large retailers created the EurepGAP (later GLOBALG.A.P.) standard to set standards across Europe.[26] By then, however, the regional standards were too entrenched to be replaced. The ironic result was that GLOBALG.A.P. became a fourth standard, specializing in produce from the developing world.[27] By 2006, the four major standards covered more than three-quarters of the food sold in supermarkets worldwide.[28]

Meanwhile, many European supermarkets began to recognize more than one standard at a time.[29] This made sense to the extent that firms did not care about the strength of standards above some threshold, or accepted that the same performance could be enforced in different ways. In 2000, a group of international retailers created the Global Food Safety Initiative (GFSI) to formalize the process.[30] By 2006, GFSI had recognized all four regional standards.[31] Three years later, GFSI-approved standards covered 65 percent of worldwide retail food sales.[32]

Politics. Supermarkets drafted the first private food standards knowing very little about their suppliers' operations. This made it hard to design cost-effective rules. The first and most obvious strategy was to ask their suppliers for advice. All of the major standards developers tried this.[33] But the suppliers were suspicious, fearing that disclosure might show anchor firms that even more onerous standards were feasible. A second, more successful strategy relied on competition: retailers announced new standards and gave a purchasing preference to suppliers that adopted them. This pressure was often intense.* Finally, GLOBALG.A.P. and GFSI adopted a third strategy by opening their organizations so that suppliers received roughly as many votes as retailers.[34] This encouraged suppliers to share information, safe in the knowledge that anchor firms could not impose unilateral burdens.

Anchor firms also used delegation to address a second issue. No matter what industry did, regulators and government were bound to accuse it of empty talk or "greenwashing." This persuaded GLOBALG.A.P.'s retailers to

---

* Retailer demands alarmed suppliers who "didn't know which way to turn because their system had been operating for 10 years with little change and they had not kept pace. Suddenly they faced the real prospect of losing a lot of business." Wellik (2012).

share still more of their power with NGOs that enjoyed enough public trust to vouch for the standards.[35]

Not surprisingly, the new organizations inherited norms and institutions from older industry-standards bodies. This included a fanatical devotion to "consensus." As one member said, "If we hadn't succeeded in obtaining a consensus we would have voted. I don't know how we did it, but we always avoided a vote."[36] The new bodies typically started by developing standards that commanded consensus in small groups of key players. The proposals were then presented to successively larger audiences, culminating in a simple yes/no vote by the full membership. Paid staff shepherded the process, making sure that the proposals were maximally transparent to members.[37]

Interactions with Government. The new private food standards preempted whatever political pressure had previously existed for government to intervene.[38] In 2002, the European Union (EU) gave business primary responsibility for food safety.[39] This made sense given the stringency of private standards, the high cost of government regulation,[40] and regulators' continued ability to influence private standards informally.[41] Congress similarly told the U.S. Food and Drug Administration to accredit private standards, though this was initially limited to imported foods.[42]

Scholars agree that private food standards created a kind of private government[43] that imposed significant costs on thousands of suppliers around the world.[44] The system currently performs more than 200,000 private audits worldwide compared to about 20,000 for FDA and "tens of thousands" by state governments.[45]

2.3. Fisheries

Government-to-government diplomacy produced broad consensus around the principle of "sustainable fisheries" in the 1980s, but little implementation. Bitterly disappointed, many NGOs began to seek alternative strategies.[46] Meanwhile, the collapse of the Grand Banks cod fishery reminded companies that their future depended on conservation.[47] In 1995, Unilever invited the World Wildlife Fund (WWF) to discuss self-regulation.[48] This led to a series of conferences in 1996–97 to turn the official consensus into a detailed principles document that could be implemented in the field.*

---

* The principles were drafted at an experts' workshop in September 1996, followed by eight meetings where stakeholders around the world were invited to comment. The series ended with a second experts' workshop that finalized the document in December 1997. Howes (2008) at p. 84; Hale (2011) at p. 309. WWF and Unilever paid for MSC's $1 million-plus start-up costs equally. May, Leadbitter, and Weber (2008) at p. 14. Participants included

Organizers' willingness to expose the principles to criticism in open meetings was instrumental in convincing observers that the document was reliably mainstream. However, Unilever's involvement still deterred environmental and social NGOs from joining.[49] This persuaded Unilever to spin off the newly formed Marine Stewardship Council (MSC) as an independent body in 1999.[50]

Joining MSC was expensive: fisheries typically spent up to $500,000 to become certified.[51] Despite this, the investment often made sense for two reasons. First, consumers in some markets were willing to pay a premium for certified fish.* This notably included high-end species like salmon: by 2006, MSC had certified more than 42 percent of the worldwide salmon catch. However, this strategy was limited to the relatively few species that command a premium.[52]

The second and much broader reason for certification concerned the "whitefish" species used in processed foods like fish sticks. These markets were dominated by large firms that seldom competed on price. They did, however, compete for market share in an industry where consumers were notoriously willing to switch brands over small differences in quality and marketing intangibles. This made claims of environmental "sustainability" a powerful sales tool. Still, there was a catch. Because Unilever could not raise the price of its products, certification costs had to be funded from existing profits. But that was only possible in markets where suppliers or anchor firms were charging above-cost prices already. This set up predictably fierce bargaining over who would pay for improvements.† Looking back, there were clear limits to how much Unilever was willing to spend. This explains, among other things, why MSC certification is nearly nonexistent among the small-margin suppliers that fish in highly competitive markets.[53]

By 2006, MSC standards covered one-third (32 percent) of the world's whitefish production and three-quarters of all salmon.[54] National fisheries similarly used sustainability to build market share for new products like

---

government scientists, activists, marine conservationists, academics, and industry. No fishermen were consulted. See Wortman (2002) at p. 78.

* Price premiums vary dramatically from species to species and are strongest for salmon. Howes (2008) at p. 87. Pollock also earn a premium in some markets. Roheim, Asche, and Santos (2011) (reporting 14.2 percent premium in politically correct London). The price for Thames herring doubled when several seafood outlets demanded MSC-certified suppliers. Prokopovych (2012) at p. A-196.

† Nassauer (2015) vividly describes how Walmart and other anchor firms demand endless concessions, which suppliers counter with equally endless threats to go out of business. This pits big firms' purchasing power against suppliers' superior knowledge of what they can afford.

New Zealand hoki, South African plaice, and Australian lobsters.[55] As of 2011, MSC either certified or was in the process of certifying 12 percent of world fishery tonnage.[56]

Unilever's leverage was focused on the roughly one hundred firms in its supply chain.[57] Instead of trying to guess which standards were feasible, the company set up procurement policies that rewarded those who found a way to offer improved practices. Specific incentives included requiring suppliers to show that they were "progressing" toward sustainability;[58] announcing a purchasing preference for MSC-certified seafood; pressing long-term suppliers to certify;[59] and setting target dates for transitioning to 100 percent sustainability.[60] By 2002, Unilever was buying the "bulk" of its fish from "sustainable" or "well managed" suppliers.[61] As in our tuna example, many fishermen, wholesalers, processors, and distributors found it cheaper to adopt the new standard for all purposes than to maintain overlapping facilities. The result is that only 10 percent of MSC-compliant fish is labeled as such.[62] The trend was further reinforced when the increased availability of compliant fish persuaded large retailers like Sainsbury's and Walmart[63] to use "sustainability" as a weapon to gain or defend market share.

Politics. MSC was originally organized around its publicly vetted principles document. In theory, this meant that all basic political and democratic choices had already been made so that enforcement could safely be delegated to apolitical experts. This technocratic vision was enshrined in MSC's constitution, which gave members little or no say over how policy was executed. Instead, the fifteen-member governing board exercised nearly complete discretion to set certification rules,[64] decide which companies could become outside certifiers, and determine which fisheries were ultimately certified.[65] Indeed, the board openly resisted calls for more democracy on the theory that it would paralyze decision-making and force the board to implement policies "it did not agree" with.[66]

Despite this formal structure, politics continued to play out each time a new fishery was certified. Most of these disputes were mediated by third-party certifiers, who used their broad discretion under the principles[67] to offer solutions that balanced the total cost to fisheries[68] against NGOs' need to show significant improvements. While the board has the last word on these deals, the process is "highly politicized,"[69] with disappointed players routinely threatening to leave MSC.[70] Despite its broad formal powers, the board's real options are limited. On the one hand, if enough NGOs walk out, MSC knows that its reputation with consumers will collapse. On the other hand, if enough fisheries depart, MSC will have nothing to sell and lose all credibility. Either way, the organization's future depends on how well the board can navigate between these disasters.

Government Interactions. MSC's standards let foreign governments and NGOs influence fisheries,[71] subject to pushback from local officials.[72] Initially, the biggest challenge came from some Scandinavian governments that resisted private regulation of their fishing fleets and pushed the United Nations' Food and Agriculture Organization to develop a replacement scheme. This led to a series of nonbinding guidelines in 2005. But MSC surprised observers by immediately embracing the document and revising its own practices where necessary.[73] The net result was to entrench MSC even more deeply as fishing's dominant certification program.[74] The following year, the dominoes began falling as Walmart and other big companies started to demand MSC-certified products. This in turn persuaded many governments to endorse MSC,[75] certify national fisheries, and subsidize private fisheries' certification costs.[76]

MSC fisheries (certified and in assessment) now account for more than 10 percent of the annual, global, wild-capture harvest, including nearly one-half of all whitefish, 40 percent of wild salmon, and nearly 20 percent of lobsters.[77] The extent to which MSC has actually improved fisheries is harder to measure, particularly since most members already had high standards before joining and sometimes use certification to resist further improvements.[78] Despite this, certification has probably produced modest increases in sustainability.[79] Meanwhile, certification remains confined to fisheries in the developed world[80] with almost no penetration of the Asian markets that consume two-thirds of the world's seafood.[81]

2.4. Coffee

U.S. diplomats organized coffee cartels to maintain Latin America's political stability throughout World War II and the Cold War.[82] However, the agreements collapsed after the fall of Communism in 1989. Coffee prices fell 40 percent over the next five years.[83] This impoverished farmers, suppressed investment, and threatened the long-run viability of coffee supplies.[84] By the early 2000s, producer income was at a thirty-year low.[85]

NGOs organized various certification schemes to stabilize prices in 1995–97.[86] However, these were limited to suppliers that sold to specialty stores whose consumers were willing to pay more for "sustainability." This only accounted for about 10 percent of growers.[87]

In 2001, the German government[88] initiated meetings with NGOs[89] and industry players Kraft, Nestlé, Sara Lee, and Tchibo.[90] The group quickly decided to launch a standard that would cover coffee sold in mainstream markets like supermarkets and restaurants.[91] But while everyone agreed on the need for higher coffee prices,[92] the big roasters equally insisted that any "major price increases" pay for better infrastructure, working conditions, and environmental practices.[93] This framed an implicit bargain by which suppliers would make new investments in return for price increases.[94]

Organizers originally hoped to house the resulting association within an intragovernmental agency called the International Coffee Organization. However, the coffee-producing nations blocked this. Undeterred, the group organized a private launch instead.[95] Governments seconded staff to the project but warned them to take a carefully neutral position[96] so that members could make their own choices.

Market Power. The group's four large anchor firms together controlled about 40 percent of worldwide coffee sales.[97] This gave them significant leverage over suppliers, particularly in Europe. However, demand for certified coffee was weak, with only one in twenty customers prepared to switch brands over social or environmental issues even at identical prices.[98] This meant that anchor firms' costs would go up without any benefit to gaining or defending market share. Their ability to extract contributions from suppliers was also limited. While some coffee chains established durable relationships with "preferred suppliers," others switched back and forth on the open market using short-term contracts.[99] Organizers tried to cover this weakness by bringing still more players into the scheme. By mid-2002, membership had expanded to include growers whose knowledge would make the standard more cost-effective[100] and NGOs who could endorse its social benefits.[101]

In January 2003, the German coffee industry trade association and the German government pledged €1 million to support development of a formal Common Code for the Coffee Community and set baseline standards for suppliers.[102] The proposal was met with widespread skepticism, and it was initially unclear how many players would participate. In the event, most of the thirty-five big coffee companies, producers, and NGOs invited to the kickoff meeting agreed to join a Steering Committee.* This allowed the project to move forward. Thereafter, organizers worked constantly to preserve and expand this membership. Drafting proceeded by consensus,[103]

---

* Founding members included Kraft, Sara Lee, Tchibo, Nestlé, Oxfam, Rainforest Alliance, Utz Kapeh, the Flanders International Cooperation Agency, and the German Organisation for Technical Cooperation. Organizers continued to recruit new Steering Committee members throughout the negotiations. For a complete list see Künkel, Fricke, and Cholakova (2008).

repeatedly forcing dissenters to choose between agreeing and walking away. This led to "many conflicts" with "almost every stakeholder group threatening to leave … at some point."* The Code was announced in September 2004.[104]

In 2005, the project announced the next step: creating a new body ("4C") to administer the Code.[105] By then, the British and Swiss governments along with the coffee industry's European trade association had pledged an additional €1 million.[106] Once again, drafting proceeded by consensus and walkout threats,[107] with final agreement emerging from a series of "We'll join if you'll join" negotiations in 2006.[108] Producers, fearing that standards would leave them worse off than before, used this leverage to write a constitution that gave them more votes than the big roasters or NGOs.[109]

4C's First General Assembly met in April 2007. Its members included seventy-one coffee manufacturers, trade associations, and producer organizations,[110] divided into separate sub-chambers for producers, industry, and NGOs.[111] The Assembly also elected a Governing Council to manage the organization and develop policy.[112] A professional staff was put in charge of drafting background documents, scheduling meetings, and retaining experts to build transparency and consensus.[113] The first 4C coffee went on sale in October 2007.[114]

By 2010, the 4C members and their de facto ally Sara Lee[115] accounted for roughly 80 percent of all certified coffee purchases worldwide.[116] While certified coffee accounted for only 8 percent of world exports in 2010,[117] this was still enough to cover 240,000 workers in twenty countries,[118] ban ten "Unacceptable Practices," and commit suppliers to making steady improvements across twenty-eight additional sustainability issues.[119] But the big anchor firms failed to follow through with purchases. As of 2009, the supply of 4C coffee was twenty times larger than actual purchases and commanded no discernible price premium.[120] Although 4C mounted initiatives to fix the problem,[121] supply was still more than twice the actual purchases in 2010 and 2011.[122] 4C announced that it was reorganizing itself around a new, "demand-driven" strategic plan in 2011.[123] The fact that suppliers had upgraded their procedures without receiving higher prices strongly suggests that they had misjudged the market and/or big roasters' willingness or ability to follow through on the promised premium.

---

* Künkel et al. (2008). Individual threats came from Oxfam, Greenpeace, and Food First Informations-und-Aktions-Netzwerk. Greenpeace and Food First subsequently carried out their threats. Id.

2.5. Forestry

Tropical deforestation became a prominent issue in the 1980s.[124] But while rich nation diplomats were able to build consensus around a concept of "sustainable forestry," efforts to implement an enforceable treaty collapsed in 1992.[125] Frustrated, the World Wildlife Fund ("WWF") turned to NGOs, small timber producers, and high-end furniture makers to organize a new Forest Stewardship Council (FSC) instead.[126] Austria and other governments also provided support as a way to circumvent trade rules that kept them from restricting timber imports directly.[127]

FSC's organizers saw the timber industry as obstructionist. They therefore adopted constitutional provisions that consigned business to a permanent minority.[128] This let FSC interpret "sustainable forestry" broadly to include ambitious biodiversity, ecology, and social-justice goals.[129] But it also meant that the organizers could not expect lumber suppliers to participate voluntarily. They therefore asked their activist allies to pressure big anchor-firm retailers into demanding FSC-compliant products. The campaign, which included more than six hundred demonstrations against Home Depot,[130] ended with a compromise in 1999. Home Depot's immediate concessions were limited to terminating vendors that violated local law[131] and promising to stop purchases from endangered regions by 2003.[132] However, the company also encouraged suppliers to compete in developing stronger sustainability standards over time. Specific incentives included a purchasing preference for FSC when price and quality were similar,[133] pressing suppliers to adopt FSC,[134] and adopting a long-term "intention" to purchase all wood products "following rules that only FSC currently meets."[135]

Home Depot's move was promptly followed by its main U.S. competitor, Lowe's. This immediately locked up roughly one-third of all U.S. home improvement stores,[136] with more home improvement chains across the U.S. and Europe soon falling into line.[137] The standard also attracted scattered adherence by large paper,[138] homebuilding,[139] publishing, and furniture manufacturing firms.[140] Demand for the new standard was further amplified when suppliers found it cheaper to adopt FSC for all purposes than to maintain redundant systems.[141] The net result was that roughly one-fourth of the U.S. market expressed "a purchasing preference" for FSC given comparable price and quality.[142]

Given that most consumers refused to pay a premium,[143] the new higher compliance standards had to be funded from existing profits. Anchor firms probably shifted many of these costs onto suppliers.[144] However, this model only worked where suppliers themselves commanded high lumber prices.

Thus, certification was nearly universal across the 27 percent of U.S. forestland owned by large timber companies like Weyerhaeuser. On the other hand, smaller producers remained outside the system, selling wood at low prices without certification.[145] This pattern was especially strong in the developing world.[146]

Results. Certification has almost certainly increased conservation efforts.[147] Additional improvements in the developing world include better health and safety measures for workers[148] and more community engagement by plantation firms.[149]

Compliance costs range from a few percent of revenue for lumber to 10 percent or so for pulpwood[150] – not too different from the estimated 19 percent "tax" that government regulation imposes on the U.S. economy,[151] although the effect is partly negated by more logging of unprotected land outside the standard.* FSC rules also generate new compliance and outside auditor expenses.[152] Like all regulation, these are wasteful to the extent that companies would have been compliant anyway, draining funds that might have otherwise been invested in physical improvements.

Backlash. Excluded from FSC, the suppliers fought back. In 1994, American lumber interests created a Sustainable Forests Initiative (SFI),[153] which reinterpreted the sustainable-forestry concept in ways that let members adjust to local conditions.[154] Despite the organizers' careful efforts to anticipate what the public wanted and suppliers would accept, roughly 5 percent of the American Forest and Paper Association's members resigned rather than accept the standard.[155] By 1999, SFI had spread to more than 85 percent of U.S. paper production.[156] It also covered more than 90 percent of the forests owned by large timber companies, although the "vast majority" of privately owned forests remain uncertified.[157]

Small-supplier opposition to FSC took a different turn in Europe, where national-level forestry organizations were already well established. These banded together to recognize each other's standards under a Continent-wide umbrella group called the Pan-European Forest Certification Council (PEFC).[158] Local governments generally supported PEFC as a counterweight to what they saw as FSC's outside interference.[159] Like SFI, PEFC argued that sustainability goals should depend on local conditions[160] and be affordable for small forests.[161] The two organizations merged into a single,

---

* Cashore and Auld (2012) at p. A-102; Brown and Zhang (2005) at pp. 2061–3. The benefits of private standards would disappear in the limiting case where every tree taken out of production on regulated land was replaced by a correspondingly vulnerable tree on unregulated land. This seems unlikely.

worldwide body in 2005.¹⁶² By then, PEFC standards covered almost twice as much acreage as FSC.¹⁶³

Internal Politics. FSC and PEFC are both governed by General Assemblies that vote on all basic documents and policies.¹⁶⁴ FSC's Assembly is split into six sub-chambers depending on members' identity ("economic," "social," or "environmental") and North/South affiliation.¹⁶⁵ Each sub-chamber, in turn, acts by "consensus," defined in the usual way to include "the absence of sustained opposition" over "substantial issues" by "any important part" of "the concerned interests."¹⁶⁶ FSC's 2012 adoption of a revised Principles document based on a three-fourths majority shows that these terms are not particularly strict in practice.¹⁶⁷ The case for PEFC is similar. While its top tier operates on a one-country-one-vote basis,¹⁶⁸ its member bodies operate by consensus.¹⁶⁹

In practice, FSC and PEFC politics both proceed against the threat of walkouts,¹⁷⁰ though the partisan nature of each group probably encourages internal homogeneity and suppresses political differences compared to a hypothetical industry-wide body. Internal conflicts are also alleviated by the substantial discretion that outside certifiers receive to interpret and apply standards flexibly to individual forests.¹⁷¹

External Politics. FSC and PEFC both understand that their credibility and survival depend on holding a significant share of the world's forests. This, in turn, means tuning standards so that they attract roughly equal amounts of supply (lumber producers) and demand (retailers). Crucially, this situation is dynamic, with each side watching and changing position in response to the other's behavior. For industry-friendly PEFC, this usually means taking steps to reassure consumers. Typical examples have included adopting a more balanced and independent governance structure,¹⁷² strengthening compliance audits,¹⁷³ adopting some rules regardless of local conditions,¹⁷⁴ and adopting FSC-style protections for workers, indigenous peoples, and local communities.¹⁷⁵ Conversely, industry-skeptical FSC has wooed suppliers by adopting new rules that let producers mix certified and uncertified wood,¹⁷⁶ reduce paperwork¹⁷⁷ and relax practices for some "small and low intensity,"¹⁷⁸ old growth (British Columbia) and plantation (U.S. South) forests.¹⁷⁹ The fact that each program criticizes the others' weaknesses and tries to provide alternatives drives "learning and adaptation,"¹⁸⁰ forcing both standards to a higher level.

Despite considerable effort, scholars' efforts to compare the PEFC and FSC standards remain preliminary. Part of the reason reflects the inherent difficulty of making global comparisons between two sets of complex and ever-changing rules.¹⁸¹ However, there is also significant uncertainty as to how well written standards carry over into on-the-ground improvements.[182] Even so, there is now widespread agreement that PEFC and FSC are converging,[183] even though "significant differences within and across the programs" remain.[184]

The downside of external politics, as Errol Meidinger has pointed out, is that rival standards bodies can persist indefinitely.[185] It follows that they will continue to fight, absorbing resources that a combined standard would use to build market share[186] and causing "polarization and confusion to purchasers of forest products."[187] Some observers have predicted that governments and activists will eventually force a merger,[188] and several governments have urged mutual recognition.[189]

Interactions with Government. Governments have tried to influence private standards through various levers, including grants, public praise, treating private standards as prima facie compliance with local law,[190] and procurement policies that favor certified lumber.[191] As of 2010, Belgium, Denmark, France, Germany, the Netherlands, the UK, China, Japan, Mexico, Norway, New Zealand, and Switzerland all had central government procurement policies.[192] After FSC and PEFC began to converge, however, officials generally began to accept both standards impartially.[193] This may have been reinforced by treaty constraints[194] and an ideological commitment to let the market decide.[195]

Within the EU, around 25 percent of timber imports are subject to sustainability requirements. Some scholars claim that government procurement has been at least as important as NGO activism or industry commitment in promoting private forestry standards.[196] Particularly in the early days, this sometimes included taking sides, as when the UK found the PEFC/SFI standard inadequate in 2004.* FSC has also stimulated debate and encouraged political organization in the developing world. Specific tactics included hosting notice-and-comment procedures outside existing agencies and iron triangles, publicizing forestry issues and developing rules that challenge captive agencies to do more.[197] Finally, many governments have used information from private standards to update their own regulations.[198]

Results. Scholars who have studied FSC and PEFC records generally agree that certification has had "important effects" on forest management,[199] including "empowering ... indigenous groups and rural communities."[200]

---

* Gulbrandsen (2014) at p. 83. The UK later relented when PEFC/SFI agreed to revise their standard. Id.

As of 2011, 9.6 percent of the world's forests were certified[201] of which two-thirds (62 percent) followed PEFC.[202] Certification is now widespread in Europe (57 percent) and North America (32 percent).[203] Additionally, the prospect of exports has encouraged some Eastern European countries to certify more than 50 percent of their forests, while Russia and Canada have certified large acreage.[204] By comparison, there has been only "limited uptake" in the tropics, where profit margins are thin and compliance typically requires costly improvements.[205]

2.6. Nanotechnology

Technologists have used ultra-small (1–100 nanometer*) particles in products like paint, flour, and beer since ancient times. Today, however, science has become vastly better at making such objects. Most of the resulting products exploit the fact that nanoparticles are more chemically reactive than bulk materials. However, this equally implies that they could pose unexpected health and environmental impacts. Governments have so far done relatively little to fill this knowledge gap, reassure the public, or develop workable standards.†

Private standards offer a natural way to fill this gap. One of the earliest efforts involved chemical company BASF, which published an in-house Code of Conduct in 2004. For the most part, the language was piously vague, for example promising to take "immediate action" when hazards arise; to present "new findings" to authorities "immediately"; and to develop a "scientifically based database" for assessing risks.[206] At the same time, this looseness was at least partly compensated by the use of very specific illustrative examples, notably including a "manufacturing guide" that requires the company to use "closed systems," respiratory filters, and chemical protective suits.[207] This hinted that the document could become more specific as new practices emerged over time.

BASF's code was meant for internal use. The following year, however, DuPont and the Environmental Defense Fund (EDF) announced plans to develop a joint framework for assessing and managing nanotechnology risk that could be "widely used by companies and other organizations."[208] Two years later, the effort culminated in the 104-page *Nano Risk Framework*,[209] which describes how companies should evaluate and manage risks where

---

* A nanometer is one-billionth of a meter. For comparison, a human hair is just under one-tenth of a millimeter, i.e., 100,000 nanometers across. Answers.com (n.d.).

† The European Commission published a vaguely worded, nonbinding Code of Conduct in 2008. European Commission (2008).

the science is uncertain. Crucially, the document endorses specific data and testing procedures, along with "reasonable worst case assumptions" and "bridging" extrapolations where knowledge is incomplete. DuPont and EDF later "pilot-tested" the Framework on three DuPont products, in one case spending $170,000.[210]

Although DuPont wanted the Framework to spread, it stopped well short of organizing fellow anchor firms on the pattern of, say, our 4C or forestry examples. Instead, it limited outreach to inviting industry and NGOs to participate in workshops, presentations, and training sessions; soliciting expert and public comment on the framework; and urging insurers to adopt the framework in making coverage decisions.[211] In the end, the available evidence suggests that the framework never spread much beyond DuPont itself.

United Kingdom. Meanwhile, British social-investment fund Insight Investment[212] tried to organize a similar industry-wide standard in the UK. It persuaded the Royal Society, the UK's leading nanotechnology trade association (Nanotechnology Industries Association (NTI)),* and various NGOs and private companies to draft a "Responsible NanoCode"[213] together. The document was finalized in May 2008 following widespread public comment. The project clearly aimed at an industry-wide standard: for example, member companies were urged to promote the document to their industry associations and suppliers.[214]

Like its American counterpart, the document featured wooly principles like "engag[ing] with … stakeholders," ensuring "high" health and safety standards, performing assessments to "minimise" risks, and adopting "responsible" sales practices.[215] However, it too listed concrete "examples of best practice,"[216] expressing the hope that these would grow over time as companies "publicly explain[ed]" how they complied.[217] The organizers hoped that this would become the "starting point for a more detailed *Benchmarking Framework*"[218] document that could be used in compliance audits, which in turn would encourage further competition to improve standards. Though originally scheduled for 2009,[219] this latter document was never written.[220]

Europe. Finally, the European Nanotechnology Trade Association (ENTA) announced plans to create a Nanotechnology Code of Conduct for European Industry in 2007. But while industry stakeholder meetings were held later that year,[221] no draft was ever published. A second, more limited initiative

---

* NTI currently represents more than one hundred nanotechnology companies. See NTI (n.d.).

was launched the following year by Swiss retail food association IG DHS, backed by anchor firm retailers Coop, Migros, and Manor.[222] Its 2008 code of conduct[223] promises that retailers will make product safety their "top priority," will only sell products considered harmless "according to the latest scientific and technical findings," and will "inform consumers openly about products that incorporate nanotechnology."[224]

2.7. Artificial DNA

Scientists first learned to make artificial DNA in arbitrary, gene-length configurations in the 1970s. However, the process was expensive and limited to relatively short molecules. For the next twenty years, academic researchers slowly improved the technology. The turning point came in 1999, when a handful of corporations began selling synthetic DNA commercially. Because the new firms served many users, they could invest more in research and development (R&D) and equipment. Ten years later, gene prices had fallen from about $5 per base pair to $0.50 on average.[225] By that point just four companies – Geneart, DNA2.0, Blue Heron, and IDT – possessed about 80 percent of the industry's installed capacity.[226] This group was overwhelmingly American,* but competed with about fifty smaller companies – mostly in the United States, Germany, and China – worldwide.[227]

Low prices meant new science. The old genetic engineering had laboriously cut and pasted DNA from existing organisms. The advent of cheap, made-to-order DNA let academic and commercial researchers skip this step, permitting many more experiments. Scientists also began dreaming of more ambitious projects like constructing complicated, machine-like organisms to make jet fuel or hunt down cancer cells. By 2000, these workers had self-identified as part of a new discipline called "synthetic biology."[228]

But there was a dark side. At least in principle, the new technology would let terrorists build simple, genetically engineered weapons without having to spend billions the way the old Soviet Union had. More ambitiously, they could also "resurrect" organisms that were either extinct or locked away in government facilities. They might even make advanced weapons with weird new capabilities like the ability to target specific ethnic groups.[229]

Government was slow to respond. Despite near panic after 9/11 and the Washington anthrax attacks, the United States produced no regulation at

---

* The partial exception, Geneart, was headquartered in Germany but maintained substantial facilities in Toronto and San Francisco. It was purchased by U.S.-based Life Technologies in 2010. Biospace.com (2010).

all until 2012 – and even then its "Guidelines" were voluntary.* The possibility of an international treaty – which would take decades to negotiate[230] – was never seriously considered. Meanwhile, the synthetic-gene industry already existed and needed guidance. This made the idea of self-governance – which had long been favored by synthetic biologists† and security professionals‡ – extremely attractive. Pressure was also building within the industry itself. First, most executives felt a social obligation to operate responsibly.[231] Second, the industry faced a slow drip of bad publicity. This included experiments that used artificial DNA to make the polio and 1918 influenza viruses[232] and newspaper claims that the industry did little or nothing to screen for safety.[233] Most importantly, large pharmaceutical firms had quietly begun asking what the artificial-DNA makers were doing to detect illicit orders and threatening to withhold business from suppliers who failed to act.§ This made perfect sense, since Big Pharma stood to make much larger profits from the new technology than the DNA makers themselves: an average "blockbuster" drug could easily earn twenty times as much revenue per year as the entire artificial-DNA industry.[234]

Most artificial-DNA manufacturers introduced some form of screening in the years after 9/11.[235] However, practices varied from company to company and some Asian firms practiced no screening at all.[236] Starting in 2004, synthetic biologists began discussing industry-wide standards to fix this.[237] Part of the impulse was practical: uniformity would stop terrorists from selectively attacking whichever company offered the fewest protections for a particular sequence.** Politically, self-regulation would reassure

---

* The US National Academy of Sciences convened a panel in 2002 that eventually called on the U.S. government to issue binding regulations. Fink Committee (2004). The recommendation was referred to the government's newly established National Science Advisory Board for Biosecurity (NSABB). NSABB (2006) at p. 8. NSABB similarly recommended that government issue binding regulations. An interagency group was established for this purpose in June 2007. As recounted below, this led to draft Guidelines in November 2009 and the final document on October 13, 2010. Eisenstein (2010) at p. 1225. Tucker (2010).

† The fact that many synthetic biologists claimed to be inspired by electronics technologies as much as biology made them especially receptive to Web precedents.

‡ Prominent bioterrorism expert Jonathan Tucker argued that formal treaties were "not a viable solution" because they would take "years to negotiate" and "be difficult to modify." This made self-regulation by suppliers "a better approach." Governments could reinforce the process by, for example, insisting that grant-funded researchers purchase all of their DNA needs from suppliers that had previously joined one or more industry standards. Tucker (2010).

§ This included several instances in which large, repeat customers went out of their way to confirm that vendors had screening programs in place. Most gene-synthesis companies experienced fewer than one such conversation per year. Maurer et al. (2009) at pp. 7–8.

** Artificial-gene companies were not the only way that terrorists could potentially fabricate gene-length DNA. However, the alternatives were comparably difficult. As executive Peer

governments and large clients that the risks were being managed[238] and might even preempt the kind of public outcry that had bedeviled GMO products in Europe.[239]

Screening Technologies. The policy choices necessarily included some mix of investigating the biology of what had been ordered and confirming the customer's bona fides. On the biology side, there were basically three options. First, human screening began by comparing customer orders against the U.S. government's exhaustive GenBank database. At this point human experts would examine the closest matches for health or security risks in the literature. This step could occasionally take PhD experts up to two hours – a significant expense in an industry where the average order cost just $10,000.[240] Second companies could use computers to compare orders against predefined lists.[241] This had the advantage of costing very little. On the other hand, only a small fraction of potential threats had ever been catalogued and this would remain true for at least a decade.[242] Third, companies could use advanced artificial intelligence to extract threat information from the literature and/or the sequence itself. But this remained a distant goal.[243]

However, even dangerous genes can have legitimate uses. This meant that biology screening was not enough – companies had to screen customers as well. This included making sure that the customer existed, had a legitimate R&D need, and had thought through any safety or biosecurity issues.[244] In practice, artificial-DNA makers could safely exclude orders for harmless genes, well-known firms, and repeat customers. But that still left a few cases in a thousand that required investigation.[245] This led to further fights over the extent to which such tasks could safely be automated.*

Industry Structure. Artificial DNA is a highly nonstandard good whose production cost can vary by 50 percent or more depending on which gene is requested.[246] This makes it hard for customers to know when they are overcharged, particularly since individual orders are too small and too urgent to solicit bids on the open market. Instead, large customers typically form relationships with small numbers of "preferred vendors" who know their needs and promise low prices.[247]

Staehler remarks, "It takes quite some effort to establish a lab for this. It is doable, but requires some skill … The risk here is medium to low." May (2010).

* DNA2.0 executives repeatedly advocated using Dun & Bradstreet credit ratings to check customer identities. The great advantage of this approach was that it was cheap and could be readily automated. The problem is that credit ratings are most useful in evaluating companies whose identities are already known. By comparison, they do very little to expose front companies and deliberate fraud. Tozzi (2009).

Meanwhile, the preferred vendors have to make repeated capital investments to keep up with technology and stay competitive. Given thin profit margins, the only way to do this is to maintain consistently high sales volumes. But in that case, losing a single anchor customer can be disastrous.* Big customers typically use this leverage to obtain the best possible price and quality terms. That said, the power is quite general and can equally be used to impose private regulation. In 2008, drugmaker AstraZeneca announced that it would only do business with suppliers "who embrace standards of ethical behaviour that are consistent with our own,"[248] specifically including private screening standards.[249]

Private Governance (Part 1): The Americans. Academic scientists' failure to adopt private standards in May 2006 (Chapter 5) led to mild press criticism. Shortly afterward, seven North American companies[250] tried to draw the sting by announcing an International Consortium for Polynucleotide Synthesis (ICPS). The new organization promised to create advanced threat-detection software, work with authorities to report suspicious orders,[251] and harmonize existing screening procedures.[252] ICPS also talked of creating a "new governance framework," a phrase that at least arguably contemplated private regulation.[253] In practice, however, ICPS focused almost entirely on conventional lobbying, building on member companies' ties to the American government.[254]

From the beginning, ICPS insisted that any solution have "modest cost" with "little or no impact on delivery times."[255] If it did not, the industry would supposedly move "overseas."[256] This framed the problem in ways that favored cheap, automated solutions, even though ICPS conceded that the required technology did not exist.[257] ICPS also opposed the common suggestion that firms pool customer data in a central repository,[258] claiming without further explanation that this would be "impractical and ineffectual."[259] Companies would pay for this ICPS-approved software through licensing fees.[260]

Private Governance (Part 2): The Europeans. In April 2007, five German companies formed the Industry Association – Synthetic Biology (IASB)[261] This led to a daylong Munich workshop in May 2008,[262] including all seven IASB members, three IGSC participants (Geneart, Integrated DNA Technologies, and Craic Computing),[263] and various security scholars.[264] The meeting was crisp and businesslike. Participants quickly agreed to

---

* Customers keep the threat credible by maintaining side relationships with rivals so that the latter know their needs and can step in at a moment's notice. Rivals understand the game and are delighted to oblige. Grushkin (2010) (quoting DNA2.0 executive Claes Gustafsson: "There's a lot of back stabbing. In fact, I'm happy to steal clients away.").

a set of work packages that they would continue to pursue for the next three years.

The centerpiece was an, private Code of Conduct that would establish an industry-wide screening standard at or near the high end of existing practice[265] and require most members to upgrade their standards in at least some respects.[266] Crucially, it required members to take the comparatively expensive step of paying human experts to compare incoming orders against GenBank.[267] Other "work packages" promised a report describing current industry screening practices*; an online platform that would let member companies share data about gene sequences (Virulence Factor Information Repository or "VIREP")†; and a Technical Biosecurity Group that would share and further develop best practices.

IASB produced its first-draft Code of Conduct that summer.[268] Succeeding iterations were widely circulated in industry and government. Initially, the three American firms that had attended the Munich meeting suggested that ICPS might join the Code‡, an idea that came to naught when ICPS disbanded later that summer. In late August, *Nature* praised the Code in an editorial.[269] This was followed by further IASB contacts with the U.S. government's Interagency Working Group, German diplomats in Berlin, and an invited presentation at the Biological Weapons Convention's States Parties Meeting in Geneva.[270] Gene-synthesis companies continued to circulate and comment on Code drafts into the first half of 2009. In July 2009, IASB announced that it would hold a meeting to finalize the draft that November in Cambridge, Massachusetts.[271]

Standards War. Shortly after IASB's announcement, the industry's two biggest gene makers, DNA2.0 and Geneart, hastily prepared their own competing proposal.[272] They revealed it at an invitation-only FBI "stakeholder's meeting" in front of roughly one hundred industry and academic biologists at the end of August.[273] Like ICPS, the firms favored a computerized approach because, they said it would be "fast," "cheap," and "replicable."[274] But it was also true that, like all existing and reasonably foreseeable lists, the proposed standard would be markedly less capable than systems that asked human screeners to examine each new order.[275]

---

* Subsequently completed as Maurer et al. (2009).
† Subsequently completed as pilot-scale software by IASB leader Markus Fischer and UC Berkeley IT professional Jason Christopher.
‡ Geneart was simultaneously a member of both IASB and ICPS. Craic specialized in writing computer programs and was not itself a gene-synthesis company. All three attended the Munich meeting.

The existence of competing standards set up a clear choice between human and list-based precautions. From the outset, both sides paid close attention to AstraZeneca's global biosafety manager, who remained carefully evenhanded while pressing both sides for more progress and, ideally, convergence on a single standard. AstraZeneca would continue to follow and sometimes participate in meetings for the rest of the year. Meanwhile, DNA2.0 and Geneart began searching for allies, quietly enlisting three American rivals in what they later called a "secret pact," before the FBI conference had even ended.[276]

In September, *Nature* reported that a "standards war" had broken out between IASB and Geneart/DNA2.0.[277] After hesitating for a few weeks, the big U.S. companies quietly withdrew their proposal* and began negotiating a new standard with Blue Heron, GenScript, and IDT.[278] Further secret discussions continued into the fall,[279] leading to yet another industry group called the International Gene Synthesis Consortium (IGSC). Once again, the big companies advocated list-based solutions.[280]

As Jonathan Tucker explained, the dueling standards set up a struggle between two very different industry segments:

> Whereas the IASB code was drafted in an open and transparent way by all firms that wished to participate, the IGSC protocol was written in secret by a self-selected group that was limited to the suppliers with the largest market share. Thus, although IGSC has urged all gene-synthesis companies to adopt its standard, only the member companies will have a say in how the screening system evolves in the future. For this reason, smaller U.S. and European firms worry that their interests and concerns will not be taken into account during the IGSC decision-making process.[281]

Both sides refused to yield. On the one hand, IGSC almost certainly assumed that its members' market dominance – they repeatedly claimed to possess 80 percent of all installed capacity and ignored the IASB standard's existence whenever they could – would attract still more adherents based on the usual tipping dynamic.[282] On the other hand, smaller members were angry at being excluded and demanded a process that would be "responsive to their concerns."[283] Geneart, which was a member of both groups, hoped until the last minute that the competing standards would find some way to merge. The standoff reached a climax of sorts when IGSC contacted IASB at the last minute asking members to call off their scheduled meeting – while still insisting that it would keep its own meetings closed to them. IASB predictably refused.

---

* Tucker (2010). The other big American companies briefly hinted that they would review the Geneart/DNA2.0 proposal "and seriously consider switching to that." Check Hayden (2009b) (quoting Blue Heron executive). This did not happen.

Cambridge. IASB held the promised meeting on November 3. The gathering had been announced since summer and was open to anyone who wished to attend. Participants ultimately included IGSC members Blue Heron and Geneart, an AstraZeneca executive, a *Nature* reporter, and several U.S. government officials.[284] Once again, the meeting was businesslike and decisive.* Probably the most significant horse-trading came from the IASB's American counterpart (the Synthetic Biology Industry Association, or Synbia), which agreed to support the standard if IASB members limited human screening to bacteria and virus genomes. The change meant that 99 percent of orders could now be screened without human intervention after all. After a short discussion, IASB leaders accepted Synbia's argument that current technology could not make weapons from other organisms in any case.

By month's end IASB members ATG:biosynthetics, Biomax Informatics AG, Entelechon, Febit Synbio, PolyQuant, and Sloning BioTechnology had all ratified the document.[285] Shortly after that, Chinese DNA makers Generay Biotech and ShineGene Molecular Biotech unexpectedly announced that they, too, would join. Presumably, they thought that the code would help them penetrate Western markets. AstraZeneca and Craic Computing praised the Code without signing it.[286] The obvious uncertainty was IGSC. Immediately after the IASB meeting, several members praised the Code, saying that they too would consider joining.† And even if they didn't, they assured the press, their own practices were similar,[287] so that any document they did write would "... come to the same thing."[288] In the event, IGSC announced its own competing document (the "Harmonized Protocol") on November 19.[289] Despite entirely new language, the Protocol appeared substantively indistinguishable from IASB's Code including, crucially, language suggesting that humans would examine all orders. This meant that the lion's share of the industry – well over 80 percent – had now endorsed human screening.‡ The continued existence of two standards did,

---

* One might have thought that putting IGSC and IASB members together in a public forum would be disruptive. In fact, the presence of reporters persuaded many to moderate their positions. In particular, there was no repeat of the FBI conference, where some synthetic biologists had taken extreme-libertarian positions against any regulation at all.

† A Geneart representative said that the company was "evaluating this to see how this will fit with our processes"; Blue Heron said it was "assessing the code"; and DNA2.0 declined comment while praising the effort. Lok (2009).

‡ Like IASB's Code, the IGSC Protocol explicitly states that companies should "screen the complete DNA sequence of every synthetic gene order ... against all entries found in one or more of the internationally coordinated sequence reference databanks (i.e., NCBI/GenBank, EBI/EMBL or DDBJ)" and that "potential pathogen[s] or toxin sequence[s]" should then receive further review "by a human expert." UGSC (2009) at p. 2. Despite this,

however, guarantee that the standards war would continue. Indeed, IGSC expressly reserved the right to amend its Protocol in the future.[290]

The Government Steps In. The main remaining uncertainty was how the U.S. government would react. So far, the Working Group had praised the IASB and IGSC codes equally without endorsing or criticizing either approach.* By late October, it had prepared its own standard.† The U.S. Department of Health and Human Services (HHS) published this draft on November 27.[291]

Strangely, the document was weaker than either of the private ones. Instead of using human screeners or even requiring lists, HHS invented still another method called "Best Match." This new protocol required artificial DNA-makers to determine whether customer orders were more similar to organisms on the U.S. government's relatively short list of "Select Agent" organisms than any other Genbank entry. If they were not, no further investigation was required.[292] The great advantage of HHS's algorithm was that, like all list-based approaches, it could be automated at little or no cost. On the other hand, Best Match was "widely considered inadequate"[293] among security scholars who argued that it was "likely to be revised."[294] This signally did not happen. Instead, the final regulation was essentially unchanged[295] except for new language conceding that non-"Select Agents" could also "pose a biosecurity threat."‡ HHS then kicked the problem back

---

some ambiguity remained. Tucker was told that the IGSC Protocol envisaged a "Regulated Pathogen Database" that included the U.S. Select Agent List and several other national lists. Additionally, IGSC members told Tucker that they only planned to use human experts in cases where "the screening software identifies a sequence associated with pathogenicity that does not encode an entire Select Agent," suggesting that non–Select Agents would be ignored entirely. Tucker (2010). Similarly, DNA2.0 told *Scientific American* that the company would indeed compare orders against GenBank – but restrict its follow-up to seeing whether "… the gene's product shows up on, say, the Select Agents and Toxins List." May (2010) (quoting Claes Gustafsson, DNA2.0's vice president of sales and marketing). Meanwhile, IDT took the opposite position, affirming that "[t]here's never a case where we would have a gene go right into production without a human being having looked at both the sequence and the prospective customer." Eisenstein (2010) at p. 1226 (quoting IDT director Robert Dawson).

* This had the unfortunate effect of dividing knowledge: while IASB and IGSC knew a great deal about the cost and feasibility of screening, only government understood the likely capabilities of rogue nations and terrorist groups. Government endorsement of one standard or the other would have brought this knowledge to bear.

† Government officials attending IASB's November 3 meeting told the author they had completed a draft several weeks earlier.

‡ Department of Health and Human Services (2010) at p. 9. HHS admitted in the accompanying FAQ that "it is not possible at this time to provide a robust database that would identify all or even most dangerous sequences." HHS (n.d.).

to industry by arguing that it was premature to regulate non-Select Agents "due to the complexity of determining pathogenicity and because research in this area is ongoing." It did, however, urge companies to develop new "best practices" to address the problem.[296]

The Dog That Didn't Bark. It would have been natural for the United States to intervene in the August–September standards war, for example by demanding that the two standards merge, or codifying their points of agreement within its own regulations. Had it done so, industry would almost certainly have acquiesced, not only for public relations reasons but because government knew so much more about terrorist capabilities.

The reasons for the government's inaction are cloudy. However, HHS later told the press that it was reluctant to impose *any* new costs on industry.* This was at least consistent with American firms' campaign for "fast" and "cheap" standards. Other contemporary observers speculated that Working Group members were not senior enough to express positions on behalf of their departments[297]; saw their task as producing a specific document instead of making policy; wanted to avoid political enmity from either side; and were skeptical that private standards would work or, if they did, expected that success would do little to advance their careers. They may also have thought that, having completed their own draft, it was simply too late to discard what they had written and start over. Certainly, it did not help that the political balance was uneven: those pushing for human screening were small, non-U.S. companies with negligible lobbying resources that hardly ever attended meetings or, in at least one case, only found out about them after the fact.†

Aftermath. By the time HHS announced its standard, more than 80 percent of the industry's installed capacity – including IGSC, IASB, and two

---

* Wadman (2009) quoted an HHS official for the proposition that "If the guidance were much more onerous than what providers are currently doing, obviously that might be of some concern." This was a strange objection given that regulation almost always increases costs compared to preexisting practices. NSABB member Stuart Levy similarly theorized that strong regulation might be bad for security since "If we deter too much, the gene-synthesis industry will go outside the United States and outside our purview, and it will come back to haunt us." Id. See also Tucker (2010) (Working Group decided that guidelines should be voluntary "partly out of concern that binding regulations would impede legitimate scientific research and put US suppliers at a disadvantage vis-à-vis their foreign competitors.") This overlooked the fact that IASB members had *voluntarily* adopted human screening, showing that even relatively small firms could afford it.

† The U.S. Government Working Group was aware of IASB's Code from September 2008 onward, while the U.S. State Department interacted with IASB on multiple occasions including the Biological Weapons Convention State Parties meeting in December 2008 and a German Foreign Office meeting in February 2009. Maurer (2012).

Chinese companies – had signed equivalent standards. It is reasonable to think that the final percentage would have been higher still and possibly unanimous if the federal government had embraced human screening. Instead, the relatively weak federal standard soon spread to industry, with many IGSC members praising Best Match as an acceptable method,[298] while other executives reportedly asked why the industry should pursue still more private standards when Washington seemed content. No company has joined the Code or Protocol since. IASB's plans to develop additional, more detailed best practices and share threat information petered out in 2012.

3

Commercial Self-Governance I: Private Power

This chapter looks for the roots of private power. Of course, power is simple enough for governments, where collective action is based on physical coercion. The puzzle for private power is that, as skeptics never tire of pointing out, "No one can be incarcerated or fined for failing to adhere to the rules."[1] Our goal is to find a stand-in mechanism that coerces without physical force.

In what follows, we will focus on markets. This is a simplification, since dissenters sometimes accede for other reasons, for example because they consider collective action legitimate[2] or want to be liked. We address these possibilities in later chapters. Despite this, starting with simple economic models offers several advantages. First, the simplification is realistic. As Prof. Maitland points out, the whole point of competitions policy is to "… isolate[] firms from one another" so that they cannot form cartels. "While we have been successful at creating truly competitive markets, we have in the process limited firms' capacity to take into consideration anything but profits."[3] Second, materialist theories tend to be clear and determinate. This will make it easier to extract unambiguous policy advice in Part IV. Finally, we will argue that our models parsimoniously explain the examples described in Chapter 2. On the principle of Occam's Razor, more complicated theories could be unnecessary.

We proceed as follows. Section 3.1 reviews the defining features of modern self-governance. Section 3.2 emphasizes how different institutions balance individual choice against collective action. We emphasize that real private governance typically falls in the broad middle ground between perfect markets and traditional government. Section 3.3 presents a microeconomic account of how anchor firms use their market power to coerce upstream suppliers. Section 3.4 extends the analysis to the case where parties repeatedly interact within semi-permanent "supply chains."

Section 3.5 asks how anchor firms are themselves constrained by consumers. It argues that oligopoly competition often makes consumers into a "shadow electorate" that constrains firms in much the same way that traditional electorates constrain public officials. Section 3.6 asks how private standards are enforced. It argues that the New Self-Governance includes much stronger enforcement incentives than earlier schemes based on voluntary self-policing or third party audits. Section 3.7 examines how public and private standards interact.

3.1. Key Facts

All theory and analysis depends on reducing complex phenomena to a manageable list of facts, patterns, and puzzles. This section justifies our choices against the recent self-governance literature and the examples described in Chapter 2.

Ecosystem. The New Self-Governance is invariably located within industries where large anchor firms[4] purchase from hordes of small upstream suppliers located around the world.[5] Many authors analyze this relationship abstractly, emphasizing anonymous markets, oligopoly, and a classical "market power" analysis in which "all-powerful global brands ... dictate commercial terms ... [to] their 'weak' and/or 'dependent' suppliers."[6] We develop this viewpoint further in Section 3.3.

At the same time, the existence of elaborate, semi-permanent supply chains shows that this cannot be the whole story.[7] This has led many scholars to focus on ongoing bilateral relations between individual suppliers and anchor firms.[8] But in that case, the usual Coasian distinction between goods manufactured internally and those obtained through arm's length trades is blurred. Since no contract can completely predict the anchor firm's needs, outside vendors end up anticipating and filling requirements much as an in-house employee would.[9] Crucially, the arrangement gives "[f]irms with large market shares" significant leverage since they "... usually have the option to source from many smaller suppliers, each of which may have few options other than doing business with the lead firm."[10] The resulting power is completely agnostic: While it is normally used to enforce price and quality, almost any behavior (better labor practices, greener manufacturing) will do.[11] We develop this approach in Section 3.4.

Finally, a few authors have begun analyzing supply chains as a command-and-control problem for the in-house bureaucracies that anchor firms have evolved to "recognize, track, and label products and services from compliant suppliers"[12] around the globe. While scholars have so far done little to develop this insight, most agree that anchor firms can monitor suppliers' conduct at least as well as governments, especially in the developing world.[13]

One Standard or Many? So far we have considered how firms exert power over individual supply chains. The question remains when and how such rules merge into industry-wide standards. Most scholars agree that the transition depends on assembling some "critical mass" of anchor firms[14] However, they have proposed many different mechanisms. These include economies of scale in monitoring and enforcement,[15] savings from shared standards development and public outreach,[16] incumbents' desire to raise compliance costs for would-be entrants,[17] lost sales for firms that refuse to join a dominant standard,[18] and the psychological argument that a single industry standard is more reassuring to consumers.[19] Other explanations invoke externalities. The most common argument is that universal standards make sense where blame is hard to attribute, so that failures by any one firm are likely to damage the entire industry.[20] In some ways, this only deepens the mystery since the same conditions notoriously encourage free-riding pathologies.[21] The fact that real firms continue to pursue collective action in these situations requires explanation.[22]

Limits on Anchor Firms. Private power is only partly about anchor firms' ability to coerce suppliers. Anchor firms must also decide when and how to use it. In practice, this discretion is limited: Nearly all scholars agree that firms are more likely to organize self-governance when downstream markets are "highly branded" so that demand depends less on objective attributes than "marketing" and a "constructed brand identity" which makes them vulnerable to "shifts in consumer preferences."[23] It follows that private standards are "most sustainable if … they establish or protect consumer demand for the brand, hedge against risks of becoming a target of activism, or reduce production costs."[24] This is a broad hint that consumer preferences can significantly constrain anchor firm behaviors in many cases. Remarkably, this is *especially* true where consumers refuse to pay more for standards-compliant products.[25] Finally, anchor firms sometimes limit their discretion voluntarily. This notably happens when they seek "significant input from other groups."[26] Anchor firm executives may also devolve power internally, agreeing to socially responsible standards that increase profitability by bolstering employee morale, increasing productivity, and improving recruiting and retention.[27]

Results. Many authors have investigated the impact of private standards using case studies, before/after statistics, and metastudies."[28] This has produced a widespread consensus that private standards really do change firm behaviors.²⁹ Whether these behaviors translate into ecological, social, and economic improvements is harder to measure, although the evidence is generally positive.³⁰ The deeper criticism is that firms are more likely to adopt standards when they are already at or near compliance. In that case, the private standard merely vouches for existing behaviors while adding compliance costs that might otherwise be spent on improving performance still further.³¹

The case studies presented in Chapter 2 confirm and extend the foregoing patterns. Box 3.1 identifies some stylized facts.

Box 3.1. Recurring Themes

The New Self-Governance requires large downstream anchor firms. Standards exist because anchor firms impose them on suppliers. This leverage derives from anchor firms' purchasing power, the existence of stable supply chains in which parties repeatedly interact, and the presence of large fixed costs that make suppliers vulnerable to business interruptions. Empirically, anchor firm power tends to be accompanied by "preferred supplier" agreements in which dominant firms trade the promise of multiple, repeat purchases in exchange for quality and price concessions. This power is normally used to enforce business terms but can equally impose private regulation and ethics codes (e.g., AstraZeneca).³²

Anchor firms are often answerable to downstream consumers. Anchor firms often demand self-governance because consumers want it. This is true even where consumers refuse to pay more for compliant products (e.g., coffee, fisheries, food). Firms often see private standards as a tool to protect market share or enter new markets (e.g., lumber, fisheries).

Suppliers have large fixed costs, face strong competition, and earn low per-unit margins. Suppliers often adopt a single standard for all purposes. This is usually necessary to avoid duplicating large fixed costs. These commonly include investments in R&D, capital, and equipment (e.g., artificial DNA), learning the needs of individual anchor firms (e.g., artificial DNA, food), and setting up supply chains (e.g., lumber, tuna). At the same time, competition between suppliers keeps per-unit margins low (food, artificial DNA). Anchor firms often create incentives that encourage suppliers to compete to adopt or invent higher standards (food, fisheries, lumber).

Anchor firms frequently delegate power in exchange for information and endorsements. Anchor firms know that suppliers possess valuable information about the feasibility and benefits of potential standards. For this reason, they often give suppliers substantial or even complete discretion to self-regulate (e.g., food, artificial DNA, nanotechnology). Anchor firms also benefit when the public understands and appreciates private governance efforts. This encourages firms to delegate power to NGOs willing to confirm that their standards are socially responsible.

Industry-wide regulation requires a critical mass of anchor firms. Industry-wide regulation sometimes follows automatically as soon as some dominant firm (e.g., StarKist, Home Depot, Unilever) adopts a standard. However, this expectation failed badly in our nanotechnology example. More usually, industry-wide rules require extended negotiations to assemble a critical mass of firms (e.g., artificial DNA, food). Our mass-market coffee example probably lies at the outer edge of what can be accomplished through political means in a relatively un-concentrated industry.

Outcomes. Private standards seldom stop at national borders. Some are regional; others global. In any given location, some standards encompass the entire market (e.g., artificial DNA) while others are limited to specific subsegments (e.g., lumber, fisheries).

3.2. Basic Concepts: Institutions, Choice, and Collective Action

Most private governance schemes focus on negotiating and administering standards. But the standards themselves can have radically different impacts. Some private standards empower individual consumers in hopes that they will make socially responsible choices. But others suppress choice in favor of collective action. In practice, what happens depends on market structure.* This section reviews the polar cases of perfect markets (individual choice) and traditional government (collective action). We then proceed to the messy middle ground where imperfect markets mix both tendencies.

Perfect Markets. Traditional microeconomic arguments hold that competitive markets are the very opposite of power. If A wants to help the

---

* More generally, real standards often mix individual empowerment and collective action with the proportions changing over time. See, e.g., Brousseau and Raynaud (2011) at pp. 65–7 (describing how voluntary standards can become mandatory and even be subsumed into government).

environment she can buy green energy from B, but this is an act of individual choice limited to her own purchases. If C wants to buy dirty coal-fired energy from D there is nothing to stop him. In practice, standards advocates often promote private standards that make markets more perfect on the theory that A will use this power to pursue socially-responsible preferences. For example, A may not know whether the energy she buys is green or not. Here, private standards can help her to spot conforming products[33] so that firms that develop "ethical goods" will receive a premium.[34] The rub, in a perfect market, is that consumers are seldom unanimous. This means that other firms will continue to offer less responsible goods at the old price.

Economists typically analyze these situations in terms of quality ladders in which goods covered by standards are priced higher than those that are not.[35] This can lead to subtle problems. Suppose, for example, that consumers initially cannot confirm which products are green. Then no consumer can trust suppliers to provide the higher standard, in which case every consumer ends up settling for the lowest quality and lowest price.* Introducing standards in this situation empowers socially-conscious consumers so that green business models become possible. This increases the number of green products on net. But the case is different where green production depends on fixed endowments. Consider, for example, the dynamics where some forests are more vulnerable to environmental damage than others. In this case empowering individual choice will encourage green consumers to buy lumber from low-vulnerability forests at a premium while non-green consumers continue to buy from high-vulnerability forests at the original price. While introducing standards will encourage consumers to switch brands, many green and non-green consumers will simply trade places. The resulting "feel good" equilibrium might not produce any net improvement at all and could even be destructive to the extent that compliance costs siphon budget that would otherwise be spent on conservation.

Additionally, green consumers will normally find it much harder to monitor compliance in some industries than others. Transparency is often especially poor where anchor firms produce intermediate goods. While some suppliers try to maintain a public reputation ("Intel Inside!"), most are anonymous.[36] Alternatively, some monopolists may be so entrenched

---

* Consider a market where consumers value quality but cannot detect defects. Then for any given market price, sellers who offer the most defective (least valuable) products earn the biggest profits. Conversely, people who own the least defective profits are deterred from selling. This reinforces consumer skepticism leading to a downward spiral in both prices and quality. Akerlof (1970).

because of market structure or patents that they can afford to ignore consumer opinion entirely.

Government. Conventional government follows a different model: Once Congress passes a green energy statute, A and B can still have their transaction, but C and D cannot. This kind of command is almost always backed, directly or indirectly, by the threat of force.

Collective action is particularly desirable where benefits increase nonlinearly with the number of firms adopting the standard. By definition, however, this means eliminating individual choice, which immediately raises questions of when the many should be able to overrule the few. So long as dissent is costless, dissenters have no reason to ever concede, so that vanishingly small minorities can block the majority. The cure for this is private power, which we broadly identify with any mechanism that makes dissent more costly.* The question remains when collective action is permissible. Conventional governments address this by adopting constitutions that specify voting rules like majority rule, supermajorities, and outright prohibitions on certain types of laws.

Middle Ground: Imperfect Markets. We have argued that conventional government depends on preventing C and D from carrying out transactions they both desire. The challenge for private power is to achieve the same result without physical force. In practice, this often turns out to be possible in imperfect markets dominated by a few large actors. Strangely, the standards that mobilize this coercion are often similar to those that empower choice in perfect markets.

In some cases, the required market imperfection relates to the standards themselves. Economists have devoted considerable effort to analyzing how consumer choice operates when the number of competing standards is limited or consumers lack sufficient resources to trust or understand the available choices. These situations almost always lead to outcomes where some consumers end up purchasing standards they disagree with.[37] More

---

* This definition is slightly less general than conventional political science definitions which stress "imposing one's will" on others. Dahl (1957) at pp. 202–3 ("My intuitive idea of power then, is something like this: *A* has power over *B* to the extent that he can get *B* to do something that *B* would not otherwise do."); cf. Büthe (2010) at p. 1 (defining "private politics" as "'attempt[s] to impose [one's] will on others' without relying on 'lawmaking or courts'"); Baron (2003) (same).The difference between this book's definition and earlier authors is that we specifically identify power with some form of coercion. This is conceptually convenient since it excludes the otherwise ambiguous case where dissenters *voluntarily* yield because they believe that the majority has prevailed through some legitimate procedure like voting.

commonly, private coercion stems from the buying power of a few large firms. The result is "mutual coercion, mutually agreed upon."[38] The possibility of "permitting businesses to coerce themselves" is particularly attractive to firms "which are prepared to incur costs but only on condition that their competitors do also."[39] This is the identical parallelism found in cartels, suggesting a deep tension with antitrust (a.k.a. "competitions") policy.

Private standards are most obviously feasible in markets where firms are able to charge a premium for compliant goods. Many authors seem to think this is the end of the story.* The following Section explains how anchor firms can impose standards even when consumer prices remain fixed.

3.3. Private Power from Markets

We now come to the central question of how the New Self-Governance extracts collective action from markets. This Section provides a first answer by analyzing industry-level supply and demand. Consistent with the literature, we find that private industry-wide standards require a "critical mass" of anchor firms. Crucially, this agreement need *not* be unanimous. We argue in Chapter 4 that this participation threshold supplies a *de facto* constitution or "voting rule" against which private politics proceeds.

3.3.1. Private Power in Markets

To see how private power works, consider a perfect market where large anchor firms purchase some product or technology from suppliers. Assume further that some (though not all) anchor firms (a) want suppliers to practice a private standard,† and further (b) insist that suppliers adopt the private standard for all purposes throughout their operations. (This last condition can be enforced either through explicit demands or else automatically through the cost penalties associated with duplicated facilities.‡) We now ask what happens when some dissenting anchor firms are happy to accept low standards. We say that the entire industry tips to the higher standard when a lone supplier operating under the lower standard would earn negative profit.[40]

---

* For an insistent but otherwise typical form of the argument that standards only matter where consumers are willing to pay a premium, *see* Wolsey (2007) (quoting Robert Reich).
† The choice could equally be between stronger and weaker standards, as in our lumber example.
‡ Economies of scale often imply that it is cheaper to manufacture all goods to the higher standard than operate two separate production lines. Examples in the coffee sector include Sara Lee and Procter & Gamble. Kolk (2013) at p. 230. We expect the effect to disappear once the supplier's market share exceeds twice the minimum efficient scale.

We now introduce the necessary market imperfection. Real firms are geared to produce a specific output defined by the underlying technology's minimum efficient scale. If they take on too much work their assets are overstretched and the cost of producing each additional unit rises. But if they take on too little business, they also become less efficient. The reason is that firms must make minimum payments just to open their doors every morning.* If production falls, these fixed costs are spread over fewer sales so that per-unit costs rise. This forces firms to raise prices, potentially losing still more business to rivals who are operating closer to their efficient scale.

Now consider private standards. Suppose that each supplier has a minimum efficient scale equal to 10 percent of global demand. Then to stay competitive, each firm needs to fill roughly 10 percent of worldwide orders. But if 95 percent of buyers insist on private standards, a dissenter must find replacement business from the remaining 5 percent. This forces it to operate at an inefficient scale so that its unit costs go up. This could still be sustainable if the industry is not very competitive and enjoys large markups. More usually, however, higher costs will force up the dissenter's prices as well. This could drive away still more customers and lead to a downward spiral.

Formal supply-and-demand models make these ideas more precise. (Figure 3.1). Readers should note that there are two possible equilibria: (a) a mixed state where suppliers who observe the standard coexist with dissenters, and (b) a pure state where dissenters earn negative profit so that they eventually leave the market or go bankrupt.[41] The dividing line corresponds to the literature's common "critical mass" intuition. According to our microeconomic model, the transition's location depends on *both* the number of anchor firms that demand a standard *and* the extent to which suppliers who adopt high standards receive a premium.

In practice, different participants will experience this critical mass differently. To suppliers, its origins will usually be mysterious, little more than the usual intuition that big firms are powerful. For them, the "tipping" to a single standard will just be one more manifestation of Adam Smith's invisible hand. But to private governance organizers critical mass will look more like a constitution – a *de facto* voting rule that specifies how many anchor firms must agree to impose an industry-wide standard. Like all constitutions, the rule by itself is not enough to ensure action. Indeed, the fact that

---

* Fixed costs can potentially come from many different sources. A partial list includes repaying previous investments in R&D, new equipment, training employees, and learning the peculiarities of individual customers.

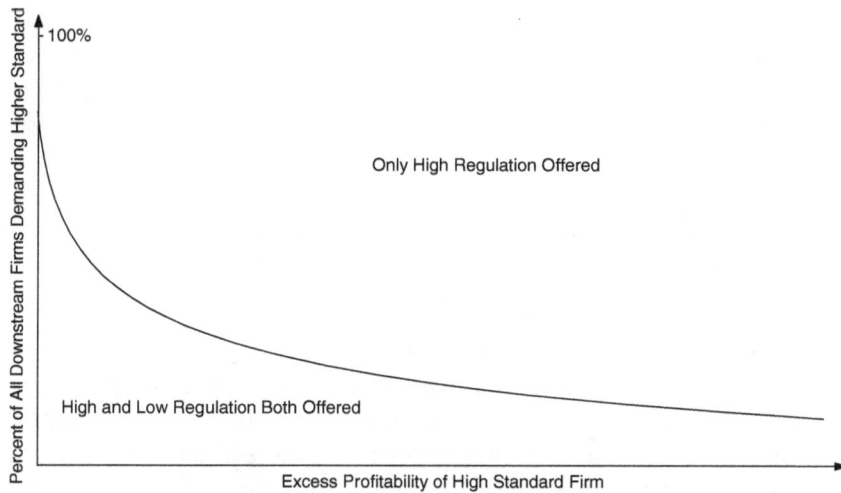

Figure 3.1 "Critical Mass" in Upstream Industries.
Source: Engelhardt and Maurer (2012).

firms have different preferences for regulation will normally mean that it is only an upper limit on how much private governance is possible. In many cases, actors will be further constrained by politics, including their inability to see that collective action is possible or negotiate alliances to achieve it. These frictions will be much reduced where power is concentrated in a few large anchor firms, as in our tuna example.

We finish our analysis by adding three remarks. First, security scholars frequently assume that profit-maximizing firms can always earn more by selling to terrorists. This proposition is wrong so long as fixed costs are large compared to any possible terrorist orders. It is hard to imagine an industry where this would not be true.

Second, we have so far assumed that compliance costs are negligible. In practice, some measures could be expensive enough that compliant firms are forced to raise prices.* This will make it easier for dissenters to stay in the market so that the "critical mass" threshold in Figure 3.1 shifts upward. It follows that security standards stringent enough to turn away even a few percent of customers can be self-defeating.†

---

* Regulation can also impose significant non-monetary costs in the form of delay, inconvenience, and canceled orders.
† The objection applies equally to public regulation in international markets where some, but not all firms are regulated. For example, Western government regulations sometimes make technology exports more expensive or inconvenient. This shelters unregulated

Finally, we have assumed a frictionless model, i.e., that buyers and sellers can readily find each other. In practice, dissenters will have to rebuild lost sales volume by persuading multiple small customers to switch vendors. This will normally take time, particularly when the price benefits of switching are small. The net effect is to shift Figure 3.1's "critical mass" threshold downward. But, this barrier is only metastable. In the long run, we expect frictions to be overcome. It is also brittle in the sense that each new dissenter reduces the value of collective action for those who remain, encouraging still more firms abandon the standard.

Delegation. So far, we have assumed that big customers exercise power directly. However, we have seen that anchor firms often share power and sometimes surrender it entirely. One reason is information. Particularly in complex fields, suppliers often know much more about how to design cost-effective standards than their customers do. But if they disclose this information, large anchor firms may decide that even costlier standards are feasible. Delegation removes this worry by giving suppliers that share information enough votes to block such outcomes. There is also a second reason to delegate: enacting a strong standard is good, but getting public recognition for it is even better. Delegating power to well-known advocacy groups convincingly shows that the standard is more than a "publicity stunt" or "greenwashing."[42] In many cases, anchor firms might not pursue a standard at all without this added reward.

In theory, these delegations can always be rescinded. That said, we should expect them to be relatively stable so long as the expanded organization continues to generate standards that benefit anchor firms on net.* So long as the delegations last, moreover, we expect delegated power to be real enough. If it were not, the recipients would never accept it in the first place. For this reason, we can be confident that any standard that includes, say, large numbers of ecologists really does improve ecology. At the same time, our argument says nothing about how the benefits are divided. Whether by luck or negotiating skill, some parties will normally receive more than others.

---

    developing world firms that could not otherwise survive. Here stringent government standards are an illusion. The real impact is to protect local firms that observe no standard at all.

*   This is only true over the long run. In the short term, we expect rational anchor firms to tolerate individually unfavorable outcomes so long as their total expected benefits remain positive.

3.3.2. Extensions

We close by mentioning three more complicated cases that anchor firm can encounter when they try to exercise power in the real markets.

Cartel Effects. Companies normally adopt private standards to reduce risk, just as they would any other fixed cost. The surprise in concentrated industries is that companies may sometimes *preserve* otherwise manageable risks for strategic reasons. To see this, recall that firms in concentrated industries have two distinct goals: (a) Maximize current profit, but also (b) deter new competitors from entering the industry and driving down prices. In principle, the second goal may sometimes be more important than the first so that incumbents deliberately make their industry less profitable to deter entry. This suggests that firms may sometimes find it in their interest to choose minimal regulation that deliberately leaves substantial risk (and its accompanying fixed costs) in place.*

Multihoming. So far we have assumed a simple binary choice between standards. However, we have seen that large food retailers, for example, often honor more than one standard at a time. This "multihoming" is perfectly rational where buyers want some minimal level of social responsibility but do not care how it is achieved or whether some suppliers exceed it.

The question remains when multihoming is profitable. Here the benefits are clear. For anchor firms, the practice drives down costs by increasing the number of suppliers that can bid on each order. For suppliers, it eliminates the need for redundant facilities and potentially offers new scale economies. That said, there are also downsides. The reason is that multiple standards bodies guarantee redundant development, monitoring, and operating costs. The resulting losses are even higher where the competing standards attack each other, so that a large fraction of each side's investment is spent canceling the other's. This recalls the usual logic of arms races, in which both sides make costly investments to little or no purpose.

---

* For a formal proof, see von Engelhardt and Maurer (2012) at pp. 12 and 20. This is consistent with our artificial DNA example, in which the large American firms uniformly advocated less regulation than their smaller European rivals. However, other explanations are possible. First, European corporate executives, like the societies they come from, could be systematically more liberal than their U.S. counterparts. Second, the history of GMO regulation in Europe suggests that European firms may be more sensitive to public backlash. Third, European genetic engineering regulations are already much higher than American ones. This suggests that high private standards would "level the playing field."

Intelligent Actor Risk. Finally, it turns out that private governance is easier to achieve for some risks than others. Legal scholars usually focus on accidents in civilian industries where risk and expected loss scale with the number of units sold. In this case, it is rational for firms to see risk as a variable cost. The same model works reasonably well for many science risks. For example, the chances that a genetically engineered organism will escape into the environment could easily depend on how many scientists use it in experiments.

That said, most national security risks follow a different pattern. Consider, for example, a rogue state that needs to purchase artificial DNA just once to make a weapon. So long as even one supplier is willing to sell the product, the number of sales to innocent buyers means nothing. Instead, the risk of a bioweapons attack depends entirely on the number of intelligent adversaries and their capabilities. But in that case suppliers face the same terrorism risk no matter how many units they produce. For them, risk has become a fixed cost. Like all fixed costs, this shifts the "critical mass" threshold in Figure 3.1 downward, reducing the number of anchor firms that must cooperate to impose an industry-wide standard on suppliers. Remarkably, then, we expect private power to be inherently easier to organize for some issues (terrorism) than others (food safety, pollution).[43]

3.4. Private Power: A Supply Chain Perspective

Section 3.3 assumed that anchors and suppliers interact in a perfect market where each transaction goes to the lowest bidder. But we saw in Section 3.1.1 that global supply chains are persistent and feature the same anchor/supplier pairings for long periods. Anecdotally, at least, such arrangements are common throughout the economy.[44] Theory suggests that they are especially likely in markets where transactions costs are high, custom products lack well-defined market prices, and it is difficult or impossible to write contracts that specify the parties' needs in advance.[45]

This Section takes these nuances seriously by analyzing relations between individual anchor firms and their suppliers as a two-person game in which the parties accumulate trust through repeat transactions. We recognize at the outset that the resulting theory is bound to predict results that closely resemble those already described in Section 3.3. That said, this redundancy is a small price to pay if our analysis offers new insights into how high fixed costs make suppliers vulnerable to leverage or builds intuition about how government can likewise intervene.

Consider, then, a two actor model in which supplier-retailer pairs regularly do business with each other. For suppliers the combination of low unit margins and high up-front investment means that they must maintain high sales volumes to survive. Anchor firms take advantage of this by offering "preferred supplier" agreements in which producers trade deep price and quality concessions for a steady flow of orders.[46] This, however, is only the beginning of the story since the anchor firm can always rescind the agreement if the supplier fails to perform. Large DNA purchasers underline the threat by making small, regular purchases from competing suppliers. This ensures plenty of second-source vendors who know their needs and can quickly step in if the preferred supplier is terminated.

For suppliers termination is catastrophic. Lacking sales volume, their unit costs skyrocket unless and until they can find a replacement customer. This cuts into profits and could force them to raise prices so that they lose even more sales. In some cases, suppliers could even become unprofitable and leave the industry.[47] Most importantly, suppliers who do find new replacement buyers will have to build up trust all over again through a long series of small transactions. This delays their return to profitability and further accentuates their punishment.*

3.5. Market Constraints on Anchor Firms

So far we have emphasized how upstream markets give anchor firms power over suppliers. But anchor firms may themselves be answerable to consumers. This section asks when and how downstream markets limit anchor firms' freedom to legislate standards for their suppliers. We consider three polar cases: Perfect competition, perfect monopoly, and oligopoly competition.

3.5.1. Perfect Markets

We start in the classical limit where competition forces firms' economic profits to zero.† Here, the usual slur that companies behave irresponsibly "because they only care about money" cannot be right. Indeed, a company that fails to invest $1 in precautions to avoid $1.05 in expected losses will soon go broke. The traditional American view (see Ch. 8) is that perfect

---

* One might think that the penalties of termination were symmetric. After all, the anchor firm must also find new partners. If these sellers are already at minimum efficient scale, we expect new orders to make them less efficient and lead to higher prices. In practice, anchor firms can almost always avoid this problem by spreading business over several suppliers.

† "Economic profit" is equal to the "accounting profit" that appears on balance sheets less opportunity costs, i.e., what the investor could have earned by investing in her second-best business opportunity. Wikipedia (n.d.).

competition is politically desirable it leaves no room for private power or discretion. Anecdotally, many scholars argue that strong competition is common in many industries across in the economy.[48]

Even so, the complaint that corporations are irresponsible contains a grain of truth: Companies in perfectly competitive markets cannot make investments to avoid externalities, i.e., harms that fall onto others. If they do, existing competitors and new entrants that refuse to make comparable investments will be able to offer lower prices that undercut them. This means that competitive markets can only address problems where society has previously attached some price tag to the issue. In some cases this reflects individual choice, as where consumers prefer to pay the added cost of compliant goods. In others, the implied price tag is supplied through regulation, litigation, or the threat of backlash.

Regulation. Regulation adds to costs by mandating behaviors that would not otherwise be sustainable in a competitive market. This necessarily increases the price of goods sold. The downside is that it also discards private information about the best way to address the threat and the expected value of losses. Furthermore, regulators must balance enforcement effort against non-compliance. Both are costly.

Probably the main weakness of regulation is that the prices it attaches to problems are only introduced *implicitly* by way of mandates. This might be justified to the extent that the problem is believed to be both substantial and hard to price. But regulation can be costly where mandated solutions prevent firms from developing smarter, more cost-effective ways to reach the same goal.

Litigation. Firms care about lawsuits. We therefore expect them to invest in precautions up to the point where additional expenditures exceed any further reduction in expected loss. However, litigation is an imperfect signal. First, corporate law usually limits liability so that losses can never exceed the violator's assets. It follows that some very expensive precautions will never be taken at all. Second, every court system is imperfect. This systematically over- or under-states the true cost of risk to the rest of the economy.[49] Third, some court systems are worse than others. This means that firms operating under weak legal regimes can offer lower prices. Given globalization, we expect anchor firms to preferentially seek out such partners, producing more of the good than society would consume if risk was priced correctly. Finally, litigation risk falls unevenly across the supply chain. Generally speaking, liability extends upstream from whichever player controlled the offending product at the time of injury. For consumer products, this usually includes the entire supply chain. But most science products, including research tools, are intermediate goods. If artificial DNA escapes from some supplier's lab, downstream customers will seldom face liability. For them, the signal might just as well not exist.

Pulling these observations together, liability rules do a reasonable (if imperfect) job of aligning *suppliers'* expected losses with society's. This encourages them to invest in socially efficient precautions.[50] However *anchor firms* are largely immune. Scholars sometimes claim that fears of regulation and political backlash can fill this gap.[51]

Backlash. The concept of backlash assumes that the political system is so imperfect that high-profile, emotive events can panic regulators into making large and dysfunctional policy changes. For example, a mass tort or scandal could invite responses that cripple or eliminate socially valuable technologies. This reaction could take many forms including outright bans, regulations that make use prohibitively expensive, or public controversies that force downstream firms to abandon formally legal technologies. Backlash can also come from insurers, which notoriously react to scandals by setting impossibly high standards, raising premiums, or withdrawing from the market.[52] The key point in all these cases is that the new regulation is genuinely excessive in the sense that expected costs greatly exceed benefits, leaving society poorer than before. Regulatory responses to genetically modified foods, chemical plants, nuclear power, and deep water oil drilling provide plausible, if controversial examples.

Unlike legal liability, backlash operates at every level of the supply chain. Furthermore, sanctions will be proportional to each firm's profit. Since the whole point of globalization is to limit *upstream* costs, we expect the largest impacts to fall on anchor firms. This is even more pronounced in science industries that make research tools (e.g., artificial DNA) that create new products. Where suppliers compete, even revolutionary technologies may have relatively thin markups. If so, suppliers' losses due to backlash will be small even if the technology disappears entirely. Downstream buyers, on the other hand, will have lost multiple products that would have earned large intellectual property or other monopoly rewards.* The paradoxical result is that anchor firms often have a bigger stake in protecting controversial technologies than the makers themselves. From their perspective, upstream firms do not do nearly enough to prevent backlash. Demanding private self-regulation is a natural way to fix this.[53] Even if prices go up, anchor firms will normally see more self-regulation as a good investment

---

* This is nicely illustrated by the pharmaceutical industry, where the value of new blockbuster drugs greatly exceeds the artificial DNA used to invent them.

up to the point where increased prices erase the incremental benefit of further risk reduction. Where different anchor firms face different risk, they will quarrel among themselves over how much self-regulation is desirable.

But none of this is socially optimal. The reason is that anchor firm incentives reflect expected corporate profits but ignore the technology's benefit to consumers.[54] The ironic result that backlash is most effective where anchor firms are able to extract most of the product's value from consumers through patents and other monopolies. Even then, backlash remains a bad bargain. This is because the threat of backlash is only effective when there is some chance it will actually happen. When it does, the irrational loss of valuable technologies more than erases the benefits of prevention.[55]

Industry-Wide Standards. The problem becomes still more complex when firms must decide whether to pursue self-regulation beyond their own supply chains. The basic point is that industry-wide standards are hard to organize. This dramatically increases organizing cost and the risk of failure. Anchor firms are unlikely to take on this burden unless the benefits of standards are non-linear so that the advantages of collective action dominate.

Collective action can also be unstable against free rider effects. Whether this happens or not depends on who receives the benefits. Generally speaking, free-riding does not occur so long the benefits are limited to those who adopt the standard. Examples include the ability to share the standard's development costs; the savings from having more suppliers compete for orders; or the credibility that widely-adopted standards gain simply by being popular.* In each of these cases firms *only* receive benefit to the extent they join the standard. This suggests a virtuous cycle in which each new participant makes the standard more cost-effective so that still more firms join. The effect will be even stronger in the common case where there are network effects or scale economies, so that the benefits that each member receives from the standard increases with its popularity. But this can also work in reverse, so that each new defector who leaves a standard devalues compliance for those who remain and tempts them to leave as well.

The case is very different where the standard's benefits are externalities, i.e., flow indiscriminately to every members of the community. The most important example is where backlash damages firms that had nothing to do with the scandal. Such situations are common where consumers and regulators find it difficult or impossible to track injury back to

---

* This is a classic economy of scale effect: The more companies adopt a standard, the more valuable the standard becomes.

individual companies.⁵⁶ This situation is particularly common for science goods, where citizens find it is hard to distinguish losses caused by irresponsible operators from an unavoidably dangerous technology. In this environment it is no longer enough for firms to worry about their own supply chain – they have to worry about their rivals as well. Externality issues are particularly important for intelligent actor threats. The reason is that improved defenses at one firm mostly shift risk to less-protected ones, leaving society's net risk the same as before.*

In all of these cases, individual decisions to participate in the industry-wide standard usually have a tiny impact on the community's overall benefits, so that each individual member receives essentially the same reward whether she joins or not. This leads to the usual free-riding logic that each member prefers to invest nothing so that the project unravels. The obvious remedy is collective action. But private collective action faces the same instabilities as a cartel: In a perfectly competitive market, no firm can charge a higher price than any other, so that firms that can offer even tiny price cuts immediately take all of the business from firms that can't. This makes standards untenable. The only solution – politely referred to as "leveling the playing field" – is to cartelize regulation so that everyone incurs the same costs.⁵⁷ This creates extraordinary incentives for members to cheat – and, just as inevitably to suspect cheating in others – which makes cartels unstable. The usual cure is to design standards for maximal transparency so that cheating is immediately detectable. For real (imperfect) markets, firms can also build trust with one another both directly or (as in our artificial DNA case) by taking advantage of trust previously accumulated within some existing industry association.⁵⁸ The problem is that such measures could equally stabilize illicit cartel agreements. We return to this problem in Chapter 8.

3.5.2. Monopoly

Perfect competition is unlikely. To the contrary: We have argued that the New Self-Governance is constructed around large anchor firms. Firms big enough to dominate suppliers will often be monopolists in downstream markets as well.

---

* Edmunds and Wheeler (2009). In principle, risk shifting could lead to an arms race dynamic in which each firm's spending cancels every other firm's leading to too much investment overall. Cartelizing security expenditures could usefully suppress this pathology. In practice, I am unaware of any instances where this cure is plausibly better than the disease.

If they want to, monopolists can ignore consumers and public opinion entirely. More precisely, their executives must decide whether to pocket their monopoly profits or spend them on policy goals like, for example, paying workers above-market wages. Presumably they make this decision based on the usual amalgam of greed, ethics, professional norms, peer pressure, ideology, and social obligation.[59] It is probably a mistake to think that any one of these motives dominates the others: Even greed encounters diminishing returns beyond a certain income. Furthermore, senior executives, particularly in science-based industries, can be quite liberal in their politics. This makes the "social and personal norms of executives" an important input to standards setting.[60] Finally, they may decide that economics and socially enlightened policies are identical, most obviously when the industry's long-term survival requires steps to preserve its raw material base.[61]

3.5.3. Middle Ground: Oligopoly and "Shadow Electorates"

Competition and monopoly are often presented as binary alternatives. But this ignores the middle ground that almost all of our examples occupy. Here, price competition is largely suppressed. But profits also depend on how many units each firm sells. This leads to a second round of *quality* competition based on offering consumers non-cash benefits. In practice, these can be material (e.g., higher quality), psychic (e.g., celebrity endorsements),[62] or policy-oriented (e.g., sustainable production). These effects were especially prominent in our fishing, coffee, and lumber examples, where actors often pursued private standards even when it was clear that consumers would refuse to pay higher prices.[63]

To see how quality competition works, recall that oligopolies, by definition, find ways to suppress price competition so that prices are "rigid," i.e., change much less than underlying costs.[64] But this does *not* imply that they are equally able to suppress *all* competition entirely. Indeed, non-price competition is usually much more difficult to suppress since (a) quality upgrades can be hard to distinguish from ordinary product differentiation, and (b) rivals must upgrade their own products to respond. This makes it hard for members to detect and punish violations, which in turn makes cheating even more tempting.[65] The net result is that non-price competition can be fierce and even drive participants' economic profits to zero.[66] This economic dynamic routinely manifests itself as an obsession with market share[67]: Because prices are similar, there is nothing to stop shoppers from switching brands overnight.

One might think that consumers would be much less sensitive to changes in policy issues or psychic attributes than to product quality. But in fact, our fishing example provides a dramatic counterexample of how policies like sustainability can become preferred tools to defend existing markets and break into new ones. Anecdotally, at least, firms' fears of consumer backlash seem to be amply justified. Despite corporate America's best efforts to avoid controversy, it is easy to find instances where anchor firms' failure to manage social issues produced steep drops in profitability and share price.[68] This implies that corporations must please shadow electorates of consumers and workers who limit company choice in much the same way that ordinary voters constrain elected officials.[69] Anecdotally at least, firms really do seem to "…live in fear of shareholder or customer backlash."[70] Worse, this backlash is often unavoidable, with firms offending part of the population no matter what they do. In these cases the best they can do is try to keep a low profile or, failing that, sell to whichever demographics support their chosen stance.*

The question remains how well these shadow electorates reflect the wider society. The good news is that the customers who buy popular brands (StarKist) or patronize national retailers (Home Depot) are nearly identical to the voters who turn out for federal elections.[71] The bad news is that purchasing decisions blur policy choices by mixing them with distantly related factors like product quality and psychic attributes. That said, these drawbacks are not terribly different from those voters in conventional elections face when they are asked to choose based on an uncertain mix of social responsibility (e.g., building new schools), individual self-interest (avoiding new taxes), and psychic values (candidate charisma). From this standpoint, neither system is unambiguously superior to the other. Instead, everything depends on the procedural details that decide how choices are presented.†

To this point we have focused almost entirely on consumer demand. However, employees can equally become shadow electors. The reason is that corporate social initiatives often improve the bottom line through

---

* See, e.g., Safdar (2017) (describing how retailer's bathroom policy simultaneously pleased liberals and angered conservatives); Enloe (2017) (describing how liberal boycott of conservative Sean Hannity led to conservative counterboycotts).

† For example, traditional electorates almost certainly possess more power where choices are presented as referenda instead of party elections. The offsetting deficit is that voters may not have the time or inclination to study these choices as thoroughly as elected representatives would.

higher morale and productivity,[72] increased ability to recruit talented workers,[73] and reduced turnover.[74] This is particularly true for Silicon Valley firms, where firms depend on happy, lightly supervised employees to be maximally creative.[75] The net result of these transactions is that employees supply price and quality improvements that "bribe" the wider shadow electorate to make different purchasing decisions – in effect, a side payment among voters. This provides yet another example of how markets give rich voters a bigger say than poor ones, though in this case wealth is measured by talent, the possibility that unhappy employees may withhold labor, and accidents of employment that give a tiny subset of workers disproportionate influence over particular goods. This dynamic presumably holds for most science-based firms.[76]

In principle, non-price competition can continue to the point where corporations spend all of their economic profit trying to please their consumers and employees.[77] In this case, management discretion shrinks to zero. This result is only slightly less stringent than traditional governments, which can ban products entirely.

Real Markets. So far, we have found it analytically convenient to consider perfect competition, monopoly, and quality competition as pure cases. However real markets almost always feature limited-but-imperfect competition that mixes all three scenarios. This suggests that real corporate elites are driven by a complex mix of liability fears, market share, and personal preferences. This situation is not too different from that facing elected officials who feel constrained to please voters but nevertheless retain significant discretion. Crucially, the balance can be inferred from market structure. Thus, we expect firms in approximately competitive markets to respond mainly to expected liability, while oligopolists also pay attention to shadow electorates, and near-monopolists more often act on their leaders' personal convictions. The balance can also change over time: most scholars agree that increased competition and that the rise of institutional investors has steadily eroded executive discretion since the 1960s.[78]

3.6. The New Enforcement

We argued in Chapter 1 that traditional industry standards' reliance on self-policing created deep incentives to cheat. This moral hazard problem disappears for the New Self-Governance, where anchor firms gain nothing – and may suffer large losses – when their suppliers evade promised standards. Naively, we expect anchor firms to enforce private standards like any other contract right. In this sense, self-governance is no different than enforcing price and quality terms.

Anecdotal evidence suggests that firms often work hard to enforce private standards. For example, Western garment retailers sometimes inspect developing world factories up to twenty-five times per year to enforce promised working conditions.[79] Similarly, a 2007 survey of 2000 large companies found that 42 percent claim to regularly assess ethics risk in their supply chains.[80] Finally, many CEOs appear to be passionate about social issues and have hired vice presidents and/or established reporting requirements to see that standards are implemented.[81] Of course, many developing country producers try to do "as little as possible," have "developed an adversarial relationship with private inspectors, and often seek to deceive them."[82] But while anchor firm executives commonly fight over how much to prioritize standards enforcement, everyone agrees that cheating can damage market share. This is at least sometimes enough for executives to follow through by terminating violators.[83]

Of course, we know that firms' appetite for enforcement is limited. For this reason, there will always be tradeoffs, with social standards competing against other contract terms like quality and cost. But it does not follow that firms will abandon private standards "when push comes to shove." Instead, the most natural guess is that a perfectly transparent market would reflect consumers' own relative priorities. Conversely, enforcement incentives in real industries must normally be discounted for the typical case where consumers find it harder to detect standards violations than changes in price or quality.

3.7. Interaction with Public Standards

So far, we have treated private power as if ordinary governments did not exist. We now ask what happens when private and public regulations compete in the same space.

Direct Conflicts. Formally, government standards are supreme.* In practice, however disfavored or banned private standards can often survive overseas. The markets picture developed in Section 3.3 allows us to analyze the problem. First, government procurement might be large enough to satisfy our voting rule and tip the market to the some rival standard. Alternatively,

---

* The supremacy of government is subject to the large qualification that private firms frequently lobby against government standards in democracies or bribe officials to change or ignore them in autocracies.

government can use its information and credibility assets to persuade some private players that the higher standard is unnecessary or will be criticized. That said, our lumber, fisheries and coffee examples show that government efforts to block private standards often fail.

Enlightened governments may also decide that parallel private standards are useful, at least for a time. Possible benefits include information that cannot be easily obtained through normal political channels, a chance to experiment with best practices, and the usefulness of international standards in paving the way for formal cross-border harmonization treaties later on.

Crowding Out: Private Standards. Government and private standards can also be inconsistent, so that participating in both requires redundant investments. Since government standard is mandatory, we expect the private sector to be crowded out.

When official and private standards compete, each side must decide when to give way. For the private sector, the decision to continue developing rules depends recursively on the anticipated public standard. Where the predicted public regulation is expected to be higher, its private counterpart becomes irrelevant and will likely be abandoned. When the expected standards are comparable, backers of a slightly higher private standard will similarly lose interest. Conversely, the private standard will often provide valuable insurance in cases where official regulation is likely to be minimal or might not be completed at all. Finally, we expect both private and public standards development to display significant inertia: Given that the cost of finishing partially completed standards usually falls over time, neither side is likely to gain much by abandoning development late in the process.

Crowding Out: Public Standards. In principle, at least, government could decide that preparing and enforcing its own standards was not worth the expense, so that it was better to leave the field to private standards entirely. Critics often object that this logic invites private bodies to self-regulate in hopes of preempting more formal (and stringent) government regulation.* In this scenario, Industry A self-regulates just enough so that regulators turn their attention to Industry B. However, this is no bad thing. If Industry A really does adopt such a strategy, then it will indeed receive less attention. But this is only because officials believe that Industry B's problems are now

---

* Büthe (2010). The objection is often raised by those trying to block private initiatives. The authors of the Responsible NanoCode acknowledge the objection at length and even apologize that "many view voluntary initiatives with suspicion" and that "similar initiatives" in the past "have been designed specifically with the intention of obviating or delaying regulation in certain areas." Responsible NanoCode (2008).

genuinely more urgent so that it is better to reallocate their resources where they can do more good. This represents a net improvement for society.[84] Relatedly, private regulation – selfish or not – will often be the *only* way to regulate minor activities, for example regulating the use of dangerous tactics in professional sports.[85]

Nor is this the end of the story. When Industry A shifts its regulatory burdens onto Industry B, Industry B can always retaliate by improving its own practices. In this case the regulators will eventually return to their original target so that any preemption is only temporary.*

---

* The logic of preemption is not limited to governments. Conroy (2007) argues that that the forestry, fisheries, apparel, and coffee industries all developed industry-wide standards to preempt activists from pursuing even harsher standards.

4

Commercial Self-Governance II: Private Politics

We argued in Chapter 2 that anchor firms can force industry-wide standards onto suppliers. At the same time, anchor firms are often themselves constrained by the need to meet price competition, avoid legal liability and backlash, and please shadow electorates. The trouble with these simple economic models is that they are completely determinate, suggesting that firms will pick the same policies every time. For this reason, they cannot be the whole story.

The missing ingredient is politics. We know from our examples that private standard setting is contested, rough-and-tumble, and above all unpredictable. To capture this truth, we must extend our model so that it permits multiple outcomes. As in Chapter 3, we begin by reviewing the main facts that any reasonable theory should respect and explain (Section 4.1 and Box 4.1). We then introduce private politics for the particularly simple case where power is held by a monopolist or cartel. Since power is concentrated, we expect each player to possess (a) ample information and negotiating resources, and (b) broad discretion to ignore consumers. In this limit, Downs's classic theory of voting blocs* predicts victory for whichever policy that commands the largest coalition. That said, a measure of unpredictability enters through the randomly selected players who exercise power (Section 4.2). We then ask what happens when oligopoly and delegation expand the number of players to the point where individual players can no longer afford to gather and process information completely, let alone negotiate coalitions for their preferred outcomes. Following Williamson, this "information impactedness"[1] regime implies

---

* Downs (1957). Downs's analysis owes an obvious debt to Hotelling's (1929) work on competition in markets where consumers have idiosyncratic preferences for some brands compared to others.

that members can no longer be sure which policies they prefer. This devolves power onto the small minority of players who possess above-average information and negotiating resources (Section 4.3). Finally, we consider cases where anchor firms are constrained by powerful but badly informed shadow electorates (Section 4.4). This leads to a private version of mass politics (Section 4.5).

4.1. Key Facts

We start by identifying key facts and puzzles from our examples and the wider literature. Box 4.1 then collects additional commonalities from our case studies.

Politics and Players. Like traditional politicians, private players almost always act from a mix of economic self-interest and ideals, principles, values, and moral aspirations.[2] They are also diverse: compared to traditional standards bodies, the New Self-Governance is much more willing to delegate power beyond the anchor firms themselves.[3] As in our food, synthetic-biology, and nanotechnology examples, these outside constituencies typically include suppliers that possess deep technical knowledge of how to design feasible, cost-effective standards and trusted third parties whose endorsements can facilitate public acceptance of whichever standard prevails.[4]

The downside of delegation is that it increases the number and diversity of players, leading to a more contested politics.[5] Once invited, outside groups can also threaten to walk out if they don't get their way. This gives them "significant power."[6] Finally, the need to keep defections to acceptable levels gives content to the New Self-Governance's vague but insistent slogan that decisions are made by "consensus."[7]

Transparency. Even where shadow electorates exist, their practical ability to monitor and punish anchor firms is limited. Scholars argue that public awareness of private standards is "modest" and varies by country and industry.[8] There are at least four reasons for this. First, anchor firms are often distant from consumers, either geographically or owing to their location in the supply chain. Second, consumers are easily confused by "multiple opaque schemes and standards,"[9] sometimes including deliberately deceptive "greenwashing." Third, shadow electors' ability to monitor and punish anchor firms depends on history. Shadow electors are particularly likely to be passive for "latent" issues that have seldom been discussed in the past, as well as issues that are "geographically, culturally, and psychologically remote."[10] Finally, activist networks may be weak or nonexistent, or activists may reasonably decide that other issues are more urgent and/or feasible.[11] Changing these conditions normally takes years.

Probably the most obvious response to these problems is to develop procedures and institutions that reduce shadow electors' costs of gathering and processing information. Traditional governments usually do this by adopting elaborate rules and procedures designed to implementing "transparency." Anchor firms similarly try to gain credibility by establishing institutions that enforce "basic democratic principles." Common examples include notice-and-comment rules, written decisions, open meetings, and formal appeal rights. Most scholars agree that these institutional innovations are not just window dressing, but have produced a "broad and continuous expansion" of participation and transparency.[12]

External Politics. New Self-Governance bodies can also develop transparency through mechanisms that owe more to Silicon Valley standards wars than traditional government. This happens more or less automatically when actors dissatisfied with existing initiatives organize competing entities on the pattern of our lumber, food, and synthetic-DNA examples.[13] This quickly leads to situations in which rival groups compete for the support of anchor firms, suppliers, consumers, and NGOs.[14] Since anchor firms can adopt more than one standard at a time,[15] new standards face relatively few obstacles getting started, provided that they can attract enough suppliers and consumers to achieve reasonable scale economies. Judging from our lumber and synthetic-biology examples, competition between standards often accelerates evolution, prompting more delegation,[16] experimentation with new institutions, and appeals procedures.[17] Competition for adherents similarly forces competing rules to converge and innovate.[18] Competition is also inclusive, rewarding programs for developing standards "… that are acceptable to the widest available audiences, thus spanning extended transnational market chains."* Crucially, this includes designing standards not just for "what people want" but also "hopefully [for] what other people will want in the future." Nor is competition limited to substantive rules. It also improves procedures so that organizations become more transparent, participatory, democratic, ambitious, and effective as time goes by,[19] while rewarding bodies that develop new information about the shadow electorate's actual preferences.[20] It follows that improvements that confer competitive advantage on any one body will quickly spread to others.[21]

---

* Meidinger (2007) at p. 51. Meidinger offers this as a partial answer to imperialism. Id. We explore the argument further in Chapter 7.

Finally, private standards can compete with public ones. Some scholars worry that private regulations will displace regulation by democratic governments, as in our food example.[22] However, private standards also make it easier for the public to see when stronger regulation is feasible, making industry capture harder to hide.[23] The fact that private standards often exceed government ones[24] hints that competition probably improves public standards on net.

4.2. The Simplest Politics: Dictators and Juntas

We have argued that the New Self-Governance is grounded in the buying power of large anchor firms.[25] This section considers the simplest case, where power is exercised by a single monopolist or cartel that declines to share power with others. Because monopolists can suppress quality competition, they are able to ignore consumer opinion and select whatever standards they like.

Formally, these systems require just one single "vote" to impose regulation on suppliers. More realistically, we expect whatever politics exists to take place entirely within the dominant firm or cartel. It follows that the total number of players will almost always be small. This has two important consequences. First, the expected benefits per player will usually be large. This means that players can afford to invest whatever time and effort is needed to learn the issues and pursue negotiation. Second, the relatively small number of players means that negotiations can be conducted through simple face-to-face or committee meetings without elaborate institutional structures. This keeps transactions costs low while minimizing misunderstandings.

The foregoing conditions suggest that players will be highly informed, skilled negotiators who are largely immune to outside influence. In this very special case, we expect accidents and misunderstandings to be minimal so that politics reliably depends on the preferences that players bring to the table.

4.2.1. Player Preferences

All politics starts with identifying the relevant players and their preferences. For monopolists, formal power belongs to the officers and directors who exercise de jure authority over the firm. However, de facto players are also possible. For example, shareholders and employees may threaten disruptions or slowdowns unless they are allowed to participate in the final

Box 4.1. Recurring Themes

Politics Proceeds by General and Selective Transfers of Information. Members must possess sufficient information to be comfortable voting for change. This gives those who possess above-average information political influence. We identify *organizers* with the private body's formal leadership. They influence members by releasing information, controlling the timing and content of votes, and identifying and organizing coalitions. *Players* outside the leadership are similarly passionate about outcomes but rarely offer coherent alternatives. Instead, they try to block standards they oppose by releasing slanted information and, occasionally, deliberately misleading arguments and assertions that make it harder for members to reach a decision. *Trusted intermediaries* are well-informed individuals that members trust to provide impartial judgments. Players may sometimes pose as trusted intermediaries to gain more influence.

Information Is Scarce and Asymmetric. Private standards bodies are not omniscient. Instead, they routinely make mistakes about public reactions (e.g., fisheries) and the expected costs and benefits of proposed standards (e.g., coffee). Organizations must sometimes revise their standards multiple times for ordinary members to receive the approximate mix of costs and benefits they originally anticipated.

Institutions Matter. The New Self-Governance has experimented with a wide variety of institutional architectures since the 1980s. These are typically modeled on public institutions (e.g., legislatures, executive agencies), industry standards bodies, and due-process norms. Strong institutions facilitate politics by reducing members' costs of gathering information and/or negotiating coalitions. Paid staff provide predigested information and foster new coalitions by exposing proposed new standards to successively larger groups (e.g., coffee, fisheries).

Politics Is Dominated by Elite Players. Most members have little budget for negotiating or gathering information and remain passive. The total number of organizers, independent players, and trusted intermediaries seldom exceeds 10–20 percent of the full membership.

Politics Proceeds by Threatening and Executing Walkouts. Anchor firms often find it beneficial to share power with outside players. Since formal votes occur at long intervals, day-to-day politics often revolves around threats to leave the coalition (e.g., coffee, fisheries). This tactic is generic to private-governance bodies despite wide differences in constitutional rules and institutions. The vaguely defined concept of "consensus" is best understood as achieving sufficient agreement to keep walkouts within acceptable limits.

External Competition Is Common. To be credible, walkout threats must occasionally be carried out. Dissenters who are suppressed in one forum commonly respond by organizing rival standards (e.g., food, lumber, artificial DNA). Unlike traditional elections, private competition can continue indefinitely. This forces competing standards to adapt and (in our examples) converge over time, following dynamics that closely resemble interbrand competition in consumer retail markets (Hotelling 1929). Standards remain viable so long as they attract enough suppliers to serve consumers and vice versa. Anchor firms commonly adopt more than one standard at a time. Competition probably accelerates the rate at which standards develop.

---

standard. Science firms, in particular, often let junior executives participate in industry-wide standards initiatives as a form of compensation or to encourage employee retention.[26] Cartels add a second layer of complexity by spreading power across multiple firms.

Each director, officer, or employee who becomes a player brings her personal policy preferences to the table. These normally include:

Profit. Greed is a powerful hypothesis. Even so, we expect it to encounter diminishing returns for very highly paid CEOs. Furthermore, profit maximization is hard work and sometimes attracts public criticism. At some point, players may decide that this is too much trouble on the principle that "[t]he best of all monopoly profits is a quiet life."[27]

Reputation and social pressure. Economists normally assume that humans prefer to mix different kinds of reward. This suggests that very highly paid executives will often be sensitive to nonfinancial incentives, including respect and admiration from peers and the broader society.

Personal values. Players often have intrinsic goals that do not depend on external rewards or audiences. These typically include personal ethics, professional norms, ideology, and perceived social obligation.[28]

In practice, leaders' preferences are not completely random. For example, large firms could earn more profit, making them systematically less willing to self-regulate. We should also expect unusually passionate actors to self-select by trying to become active players. Their passion will often reflect comparatively extreme views for or against regulation.

4.2.2. Coalitions

So long as players have abundant information and negotiating budgets, we expect them to converge on whichever outcome (a) includes enough votes to dictate a standard and (b) offers voters the closest match to their preferred policies compared to every other feasible alternative. This logic quickly leads to Downs's (1957) classic theory of coalitions.* As Downs himself emphasized, the result is a lowest-common-denominator dynamic in which regulation is set by the least-enthusiastic member of the coalition.† When market structure requires supermajorities, we expect still weaker standards. These protect the minority against actions with which they disagree but also disenfranchise the majority.

Finally, Downs assumed that each candidate policy's costs and benefits were fixed. However, we argued in Chapter 3 that net benefits can grow when a standard becomes popular. This leads to a rich-get-richer dynamic in which small differences in initial popularity are steadily amplified.‡ To the extent that these early differences originate in unpredictable accidents and differences in political skill, we expect outcomes to approximate a random lottery over players' policy preferences.

4.2.3. Delegation, Consensus, and the Politics of Walkout

The power of anchor firms is guaranteed by economics – nothing that the players do or say can keep them from taking back power whenever they want. But in the meantime, they can often obtain more benefits by sharing some or even all of their power with others. Indeed, many projects may not be worth pursuing otherwise. Our examples suggest that this usually occurs for three reasons:

---

* As Downs himself noted, simple coalition politics can become more complex where the existence of multiple issues opens the door to "logrolling" strategies. Suppose, for example, that Firm 1 cares intensely about Issue A but not B. Then it can often improve its lot by trading votes with Firm 2, which cares about B more than A. The result is that Issues A and B are both more likely to prevail than they were before. The catch is that science-governance issues are often presented one at a time. That said, Chapter 2 documents some important counterexamples, notably including forestry regulations that conflate a wide range of ecology, labor, and economic development issues.
† We neglect Downs's discussion of multiparty systems, which finds no analog in our examples.
‡ Strictly speaking, network effects are inconsistent with our perfect-information and negotiation assumptions, which would allow players locked in an inferior standard to immediately move to a better one. We implicitly assume that negotiating costs, though minor, are large enough to block simultaneous coordinated action by large numbers of players.

Asymmetric information. Suppliers almost always know more about their own internal operations than customers do, and this is especially true for high technology and science-based industries. But in that case, anchor firms cannot know which standards are feasible, let alone how much risk can be reduced per dollar invested. In extreme cases like our artificial-DNA example, they may not be able to draft any standard at all. This limits them to choosing between whatever proposals upstream firms decide to offer.

Credibility. Companies that behave virtuously want consumers to know and appreciate that fact. This encourages anchor firms to share power with NGO partners who can vouch for their conduct. But NGO leaders are themselves constrained by the need to please rank-and-file members and large donors. This expands the shadow electorate still further.[29]

Legitimacy. Even where private power exists, there is no reason to think that market-enforced "constitutions" are desirable on normative grounds. Anchor firms may sometimes rewrite private constitutions to allocate power differently in order to persuade government and the general public that its standards are legitimate. SFI's extensive reforms in our forestry example offer a particularly dramatic example.

The problem with delegation is that there is no practical way for either party to enforce the agreement. Instead, each new private standard provides an opportunity for anchor firms and delegees to renegotiate. This explains the ubiquity of walkout threats in our examples despite wildly different institutional architectures. Walkout threats also give content to the otherwise mysterious concept of nonunanimous "consensus." Almost every significant standard will lead to a few walkouts. So long as the number stays below some acceptable level, dissenters can be safely ignored. But if large numbers leave, the expected benefits of delegation disappear, so that anchor firms may decide that self-governance is no longer worth the effort.

Players who decide not to walkout plainly believe that they can exert more power from within their organization than they could by leaving. But in that case, their leverage within the organization implicitly depends on what they could achieve by petitioning the government, the courts, and public opinion. In this sense, we expect members' private power to be allocated according to their influence in traditional institutions.

Given that delegation can be rescinded, anchor firms must continually revisit how well power sharing serves their interests. Presumably, they will continue to delegate so long as they expect the organization's private standards to deliver net benefits into the (discounted) future.* It follows that self-governance bodies will never produce outcomes that deliver negative benefits to anchor firms over the long run. But this objection is not fatal so long as the organization helps other players reach goals they could never achieve on their own. So long as the delegation endures, it is reasonable to think that a standards body that includes, say, a well-known environmentalist group, really does benefit the environment.

4.2.4. Social Welfare

We have argued that private power approximates arbitrary rule by individual dictators or juntas. The normative question is whether this might still be better than no private governance at all. One obvious argument is that most business executives, particularly in science, resemble the broader society. This suggests that private standards will reflect mainstream and even liberal political preferences on average. Despite this, we will see in Chapter 8 that American courts have traditionally rejected such arguments on the ground that unconstrained private power is dangerous no matter who holds it. The objection can be formalized by noting that a monopoly enforces the preferences of a single decision-maker selected at random. For this reason, policies will almost always be further removed from "mainstream" opinion than a Downsian coalition system that drives players to find and converge on middle-of-the-road solutions.

The counterargument is that doing nothing is also costly. By definition, markets set every externality equal to zero. From this standpoint, adding a weaker-than-mainstream standard is still preferable to doing nothing at all. In this sense, at least, dictators and juntas can only improve matters.†

4.3. Oligopolies: Internal Politics with Limited Information

Section 4.2 assumed a verify simple politics where players possess unlimited budgets to collect information and negotiate. But in fact, most of our examples like coffee, fisheries, and lumber follow a very different pattern. We now turn to the common "information impactedness" case where a large majority of members (a) do not know which proposed coalition best matches their

---

* To some extent, firms can commit not to rescind their delegations. For example, an anchor retailer that publicly embraces the Sierra Club in Year 1 and severs all ties in Year 2 is much more vulnerable to criticism than one that had no relation at all.

† Given the usual prejudice about commercial actors, we ignore the possibility that private power will generate net negative social value through over-regulation.

preferred potential policy outcome, and (b) lack resources to organize new coalitions on their own. Impactedness immediately implies that many different potential policy outcomes are stable. Following Williamson, we assume that players try to persuade members by providing additional information.[30] This information can be unbiased, slanted, or deliberately designed to confuse. The current section applies these ideas to the internal politics within governance bodies. Section 4.4 extends the concept to include external politics and shadow electorates.

4.3.1. Participants: Leaders, Players, and Members

Given that information is costly, we expect only the most committed members to engage in politics. For convenience, we stylize these groups in descending order of commitment, resources, and effort invested.

Organizers. Organizers hold formal positions in the governance body and work to accomplish its goals. These centrally include persuading members to support new rules. Organizers do this by providing information, either indiscriminately or in a deliberately slanted way. The former has the advantage of building trust in a system where members hardly ever have enough information to very claims for themselves. This is crucial where members prefer deadlock to adopting new rules in the face of uncertainty.

Leadership requires large personal commitments of time and resources. Since organizers are rarely paid, the role usually falls to those who are especially passionate about some discrete issue (e.g., biosecurity) or who want to build a particular project (e.g., a mutations database).

Players. Private politics also includes non-organizers. These players typically invest less effort than organizers but more than the average member. Furthermore, they invariably dominate public discussions.

Like organizers, we expect players to influence members by supplying information. However, they normally have fewer resources. Following Williamson, they will often find it cheaper to inject deliberately confusing arguments that create "information impactedness." This makes it costlier for members to process information and increases the chances of deadlock.

Trusted Intermediaries. By definition, organizers and players pursue specific political agendas. However, other well-informed actors could be neutral. Paradoxically, this neutrality can make them influential. Given the high cost of collecting and processing information, members may decide that it is better to trust those who know more, provided they have no visible axe to grind.* The difficulty, of course, is that trusted intermediaries may have undisclosed biases. This makes it tempting for trusted intermediaries to slant their advice and become players in their own right.

Members. The great majority of members are comparatively passive and uninformed. While voting will eventually give them the last word, they are only consulted at long intervals. In the meantime, their opinions (by definition) reflect relatively little thought or information or even passion. It follows that their preferences will often be lightly held and changeable. This makes it almost impossible for organizers and players to count heads between votes.

In principle, we expect organizers, players, and trusted intermediaries to compete to supply information. This could reduce members' costs to the point where they become fully informed. At this point, Downsian politics would reemerge so that outcomes no longer depended on accident or political skill. In practice, this seldom happens. This is particularly true for complex issues and/or organizations that are too large to resolve issues through simple face-to-face interactions.[31]

4.3.2. Private Politics: Strategies and Tactics

We have argued that high information costs limit active politics to a comparatively small subset of leaders, players, and trusted intermediaries. Each group typically possesses different resources and pursues different strategies.

Leaders. Leaders can typically call on more resources than other players. Their strategies include:

> *Education.* Members must possess enough information to be comfortable saying "yes" to a standard. For this reason, organizers spend much of their time trying to reduce members' information costs. Typical methods include mass and bilateral e-mails, meeting presentations, one-on-one discussions, and hiring staff to compile written briefing materials. However, members will demand even more information before saying "yes" if they think the disclosures are biased. All else equal, persuading members is cheapest where leaders release information evenhandedly according to demonstrably transparent procedures.

---

* Trusted intermediaries may also have openly disclosed biases, for example, environmental organizations' preference for "green" solutions regardless of cost. Provided that these biases are fully disclosed, members can still extract information from them.

Coalition building. Because of imperfect information, members are normally prepared to accept a wide range of standards. In order to prevail, organizers must assemble a coalition around one of these candidates. On the pattern of our coffee example, this often involves an iterative process of suggesting and revising proposed standards for progressively larger subgroups. The question remains which standard organizers will select from all those that would theoretically be feasible. We expect rational leaders to pick candidates that strike a balance between standards that offer the least political risk of failure and their own personal preference for some standards compared to others.

Agenda setting. Leaders control which documents are drafted, when they are debated, and when votes are held. This suggests that they can adjust the timing and content of votes to increase the chance of winning. Because members need and expect leaders to keep order, organizers will normally possess the usual chairperson's discretion to slant proceedings against those who oppose them.

Finally, it was pointed out in Chapter 1 that members often take leaders' endorsements as evidence that the favored standard will eventually prevail. This expectation can become self-fulfilling, elevating leaders who have often prevailed in the past into influential "philosopher-kings."

Players. Players can hardly ever match organizers' resources for providing education, building coalitions, or setting agendas. Despite this, they can pursue at least three tactics to pursue their interests:

Competing standards. In theory, players could advocate their own, alternative standards. This would require them to persuade members by providing information in competition with the organizers. On the usual democratic logic, disclosures on one side would then force counter-disclosures on the other so that more information became available on net. That said, our examples do not contain a single instance where this happened within a single, community-wide initiative. Players do, however, commonly leave their organizations to form rival bodies. We consider this "external politics" in Section 4.5.

Coalition politics. We have already said that players can threaten walkouts unless and until organizers meet their demands. However, limited information implies that both sides may sometimes miscalculate. Leaders may think that players are bluffing, or underestimate the extent to which a walkout would cripple the eventual standard. Conversely, players may threaten a walkout so strenuously that backing down would permanently erode their influence. Or they may carry out the threat only to find that the organization can get along without them.

Forcing deadlock. Williamson points out that it is generally cheaper to confuse members by raising their information costs than to inform and persuade them. It follows that players with inferior resources can sometimes force stalemates. Institutions and norms that favor orderly and transparent debate can help suppress such tactics, making this outcome at least marginally less likely.

This is as far as theory can take us. Because information is always imperfect, just knowing the relative resources of, say, leaders and players is only partly determinative. Political skill and small events – a chance misunderstanding, a badly turned phrase – also matter. At some level, private politics will always be unpredictable.

4.4. Oligopolies: Mass Politics and Shadow Electorates

The internal politics of standards bodies is only part of the story. In many cases, private standards must also satisfy consumers. This section explores how private governance tries to do this.

4.4.1. Shadow Electorates

The preceding section imagined politics as a contest among leaders, players, and ordinary members. We now extend our list of actors to include a shadow electorate of consumers.

Policy Preferences. As before, we begin with shadow electors' preferences. In practice, these closely overlap the actual electorate. For example, the number of Americans who shop at Target is roughly the same as those who vote in midterm elections.[32] This already guarantees similar preferences. That said, our shadow electorate is bound to overrepresent some voters compared to others. The most obvious difference is that private "votes" depend on income, so that voters earning $150,000 per year naively possess ten times as much influence as those at the poverty level earning $15,000. In fact, things are not quite so bad since differences in consumption are only about 40 percent as large as income inequality.[33] The difference is further diluted to the extent that rich citizens spend less at mass outlets (e.g., Home Depot) compared to artisans (e.g., Maserati) that exercise negligible purchasing power in global markets.

The gap between shadow electors and voters is even narrower if we accept that the poor only vote in traditional elections about half as often as the rich.[34] However, this argument is tricky, since the right to vote is more important than its actual exercise. This is especially true if nonvoters rationally decide to abstain, i.e., let active electors decide on their behalf. This hypothesis is at least consistent with the observation that nonvoters are "virtually a carbon-copy of the electorate."*

Structured Choices. The power of mass electorates to set policy invariably depends on the order in which choices are presented and bundled together. The question remains what an ideal democratic rule would look like. Probably the most natural suggestion is to let electors vote on *both* the private *and* the social costs of specific activities and nothing else. Decisions to buy individual products do this naturally. The result is similar to the referendum idea in traditional politics where voters who endorse, say, a school bond issue immediately know that they will pay higher taxes as a result.

We might also object that commercial marketing is often irrational, so that shadow electors are bombarded with psychological or symbolic arguments that have nothing to do with objective facts. Suffice it to say that this is not obviously worse than the situation in traditional elections where candidates benefit from intangibles like charisma.[35] Whether voters are more irrational in the case of candidates or products is an open question. It is reasonable to think that Nike shoe buyers are less sober than Michael Dukakis supporters but more critical than JFK enthusiasts.

4.4.2. Mass Politics

So far we have assumed that the shadow electorate can monitor and punish organizations that deviate from its preferred outcomes. But members of the public have limited time, energy, and attention spans. It follows that the electorate's actual power depends on how well institutions reduce the cost of acquiring and processing information. Whether this happens or not depends on three conditions.

First, some actor or group must be willing to invest in the hard work of education. Beyond the already substantial cost of marshalling facts and inventing arguments, activists must also complete against each other. This is because newspapers and TV news have far too little space to report every standard and injustice. Whether a particular issue breaks through also depends on the

---

* Wolfinger and Rosenstone (1980) at p. 109. The main exception is that voter and nonvoter preferences noticeably diverge for big government and redistribution issues. Leighley and Nagler (2013). These are precisely the topics where private governance has nothing to say.

number and skill of activists – itself a kind of human capital inherited from previous campaigns.[36]

Second, conditions must favor communication. We have already argued that firms that manufacture intermediate goods (e.g., Intel or Boeing) are less visible to the public. This increases communication costs. Conversely, it is normally easier for activists to communicate issues that have been raised in the past and can draw on an established public narrative.

Finally, consumers must believe that intervention is both possible and necessary. This is partly a practical question. The further removed an anchor firm is from the public, the more indirect its punishment becomes. Even if they wanted to, shadow electors would find the idea of punishing Boeing by withholding business from, say, United Airlines, inherently uncertain.*

Beyond this, voters may rationally defer to representatives who can be trusted to implement the solutions that they themselves would arrive at *given unlimited time and study*. This empowers leaders to pursue solutions less for their current popularity than because they "anticipate broader public values" and will eventually be "acceptable to the broadest range of people moving into the future.[37] In traditional governments this means electing representatives whom voters trust as individuals. But this choice is unavailable to private legislators who are almost always self-selected. This shifts the trust interaction from individuals to organizations (e.g., the Sierra Club) that already enjoy public trust and, eventually, the standards body's executive leadership to the extent that it is able to build trust through successive transactions on the philosopher-king principle.

4.4.3. Institutions

Private governance bodies have evolved various institutions to track public opinion. Most involve private legislatures. Absent mass elections, these cannot possibly "represent" the voters in the conventional way. Despite this, our examples suggest several reasons why such bodies can provide valuable information about sentiment in the wider society.

> *Stochastic representation.* Because private legislatures are self-selected, they will almost always under- and overrepresent particular viewpoints. Despite this, it is reasonable to think that large, open memberships will guarantee that every view is represented at least once and,

---

* See Chapter 10.

for mainstream views, several times.* It follows that a sufficiently large supermajority within the private legislature would likely command majority support among the general public as well.

Measurement. Private legislatures give dissenters a chance to speak up. This provides a tripwire for executives who have not encountered these views in the wider society. Here, the MFC's decision to create a deliberately powerless advisory chamber is instructive. Despite their formal irrelevance, advisory members could express themselves to MFC's leadership and even conduct walkouts to demonstrate their outrage.†

Signaling and deterrence. Finally, legislatures serve a signaling function by showing the outside world that a political consensus exists. If a private body can command agreement from hundreds of people – none of whom is particularly close to the organizers – it becomes much more plausible to think that the standard will find similar support among the general public. If executives are nevertheless willing to submit their agenda to a vote, they must really believe that the rest of us would, given enough time and study, agree with them. This provides a convincing signal to both the public and would-be activists that political challenges are unlikely to succeed. This is a fortiori true where, as in our W3C example, organizers have repeatedly prevailed in the past.

4.5. External Politics

The New Self-Governance's biggest innovation compared to traditional politics is that it allows multiple private "governments" to coexist and compete indefinitely. It would be surprising if this freedom did not feature distinct costs and benefits compared to the traditional case in which politics takes place within a single industry-wide standards body. This section reviews some theoretical reasons why external politics might sometimes be able to mobilize more effort or reveal certain types of information more efficiently than conventional institutions that develop information through deliberation and debate.

---

* We ignore cases like our forestry example where membership deliberately excludes certain classes of society.
† MFC also tried to gauge public opinion by nonlegislative methods. These included (a) outsourcing politics to third-party certifiers charged with finding standards that would simultaneously appease both industry and activists, and (b) exposing proposed rules to criticism at ad hoc meetings that included both experts and the general public.

4.5.1. Relative Effort

Probably the most basic way to analyze the differences between internal and external politics concerns how much effort and resources we expect each method to mobilize on net.

We start with the positive side of the ledger. Our forestry and artificial-DNA examples suggest that competition often accelerates the development and convergence of standards. To progress further, we need to articulate some theoretical reason why competition might do this. We argued in Chapter 3 that competition is fiercest when consumers readily switch brands over small differences in price or quality. In extreme cases, this logic leads to the familiar arms-race dynamic in which small improvements by one side immediately erase the other's investment in security. This leads to the standard pathology in which both sides invest frantically and neither receives any lasting benefit. The standard solution to this trap is cartelization, i.e., negotiating treaties that suppress further arms purchases that promise no benefit to either side.*

Consider, then, a community whose members are divided into two groups, each of which has a radically different vision of what the private standard should look like. In the **internal politics** case where there is only one standards body, both groups must agree to a compromise standard *before* it is implemented in the market. This replicates the basic arms-control-treaty logic of implementing the outcome expected from competition without actually competing. The difference for **external politics** is that there is no chance to negotiate in advance. Instead, each side simply implements its standard and lets the market decide. This leads to the usual expensive cycle of moves and countermoves. While this competition is probably somewhat muted in life,† the underlying dynamic still forces the parties to expend more effort than they would under a single, community-wide initiative. This extra effort might or might not be desirable. The benefit is that standards will arrive faster, be revised more often, and continue to be refined further into diminishing returns than they would be under internal politics. The balance of this section looks at the cost side of the ledger.

Duplication and Waste. External competition requires society to replace a single shared enterprise with two competitors. This can entail significant costs:

---

* For a formal analysis see, e.g., Kydd (2015) at pp. 37–44.
† We have already said that our model depends on consumers' willingness to switch standards over small differences. In practice, this is almost certainly moderated by (a) consumers' very limited information budgets, which imply that many improvements go unnoticed, and (b) the existence of semipermanent supply-chain relations between firms, which would be disrupted by incessant switches from one standard to the other.

Duplication. Dueling standards bodies necessarily imply duplicated development, marketing, and operating costs. The proliferation of standards also forces customers to invest in duplicate learning and may limit the number of suppliers that can bid on any particular order.

Waste. Our forestry example shows that competition often leads to standards denigrating each other. This leads to the classic arms-race dynamic in which each side's investment goes mainly toward canceling the other's efforts with no net benefit.[38]

Confusion. Standards war theorists often assume that consumers have unlimited budgets to gather and process information. This is unlikely. Some scholars argue that the existence of multiple standards deepens consumer confusion. This is especially likely when some competitors make deliberately misleading claims ("greenwashing").[39]

Intuitively, we expect these costs to reduce but not eliminate the benefit of developing external standards. After all, members would not compete at all unless they thought that the differences between rival standards were worth fighting over. That said, it is easy to imagine cases where frictions might delay the emergence or disappearance of external politics compared to our simple model.

Is External Politics Undersupplied? In theory there is nothing to stop members from transitioning from a single organization (internal politics) to multiple competing ones (external politics) as soon as there are enough suppliers and consumers for a new standard to achieve scale economies. In practice, we expect organizational frictions to interfere. That said, our forestry and artificial-DNA examples suggests that these barriers can be readily overcome. This seems logical in commercial settings where the choice between standards often translates into large differences in profitability.*

The Graceful-Exit Problem. One might expect rival bodies that saw their standards converging to realize that any remaining differences were no longer with the costs of continued competition. Following the logic of arms-control treaties, both would then realize that they were better off negotiating a merger. The striking thing about our forestry and synthetic-DNA examples is that this signally does *not* happen. Once established, standards bodies seem to grind on indefinitely. One can imagine at least three possible reasons for this.

---

* Strikingly, our academic examples do not include a single example of external competition. This may be because organizational frictions and/or fears of "splitting the community" are more important in these settings. See Chapter 6.

Careerism. Institutions create careers for the individuals who run them. Regardless of institutional interest, the leaders of competing standards bodies will seldom gain much from a merger, especially when total spending declines.

Deterrence. Converged standards are dynamic: if one body were to suddenly disappear, the other would immediately revert to more extreme standards. This suggests that rival standards bodies maintain deterrence simply by existing. That said, even a disbanded standards body can be reconstituted when needed, especially since we have argued that the costs of organizing are fairly modest.

Trust. We will argue in Chapter 6 that the ability to negotiate transactions depends on trust built up gradually over multiple repeat transactions. Internal politics creates many opportunities for this. External politics, by comparison, offers hardly any interactions short of negotiating a global agreement to suppress competition. While theoretically possible, this kind of merger agreement could be a bridge too far.

In practice, it is reasonable to think that all three explanations have some validity. If so, rival entities will be slow to disband unless and until a critical mass of anchor firms decides that competition is no longer in their interest. This could happen because there is little left to learn from further experiments or because any remaining differences between competing standards are outweighed by wasteful duplication and lost scale economies.

4.5.2. Institutional Advantages

So far we have focused on the comparative ability of internal and external politics to mobilize (and waste) effort. However, this is only part of the story since effort will almost certainly be more effectual in some contexts than others. We argue here that the relative advantages of internal and external competition depend on the type of facts to be developed.

Fact Discovery. This book has emphasized that the efficiency and predictability of private politics depends on members' – and in some cases shadow electors' – ability to obtain and process factual information. Conventional governments typically do this using committees, debates, and hearings. When this happens, we expect politics to unfold with each organizer and player strategically disclosing whatever information supports his or her preferred outcome. This is less risky than it sounds, since competition will equally create incentives to fill in skewed disclosures by others. The more dangerous possibility, emphasized by Williamson, is that players may deliberately create confusion ("information impactedness") so that members are reluctant to vote for any standard at all.[40] But this is not the only possibility. As Joseph Farrell and Garth Saloner point out, Silicon Valley–style standards wars often offer a dramatically different alternative in which each side is required not just to describe and argue for its proposal but to let members actually see and experience it.[41] Competition then proceeds when members vote with their feet by adopting whichever standard(s) appeals to them so that rules must constantly evolve to stay competitive.

From the Farrell/Saloner standpoint, the importance of external politics is that it brings Silicon Valley methods to politics. Meidinger has similarly argued that external politics offers a plausibly superior mechanism for delivering better substantive standards that "work on the ground."[42] The question is how useful these are. The answer is that traditional hearings and debate are probably the better way to determine relatively discrete and objective facts, for example, "How many fish were harvested last year." But this is not generally the kind of information that members need to cast votes. More typically, members would prefer to know *predictive facts*, i.e. what will happen if a standard is implemented. While these can sometimes be demonstrated through logical argument, members would have to invest extensive effort to verify them, including, in many cases, mastering specialized economic or scientific learning. This is seldom possible within the average member's limited information budget. In these circumstances, external competition provides a much more attractive way to experience each standard and its limitations.

The most extreme version of the predictive-facts problem involves information that is literally unknowable. In this case, progress can *only* come if members are willing to experiment with standards knowing that most will fail. In these situations, members will not even try to implement a new standard unless they are confident that further trial-and-error iterations will be carried out. We discuss this extreme case in Box 4.2.[43]

Finally, the advantages of external politics will normally be reduced where we expect standards wars to show "tipping behaviors," so that early accidents can lock in inferior standards. In this case, the differences between the winning and also-ran standards will depend on history as much as logic, suggesting that a suitably intelligent committee could make better choices.

Political Discovery. All private standards bodies develop policies in two stages. In the first, organizers discover and build consensus around proposed standards that will be acceptable to a suitable majority of members. This already generates information by showing the feasibility of coalitions

Box 4.2. Trial-and-Error Governance: 4C Coffee Revisited

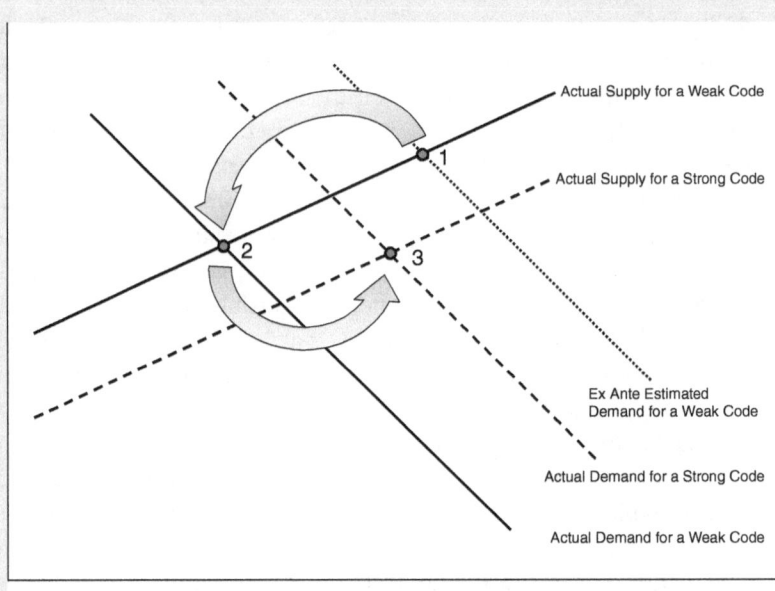

Figure 4.1 Trial-and-Error Standards Development.

Democratic choice conventionally assumes that members know how the benefits and burdens of each proposed policy would be allocated across members. However, in many cases these facts can only be determined by experiment. Recalling our examples, no one could know with any certainty how new sustainability measures, say, would affect fish populations or whether new coffee standards would increase growers' profits enough to pay for themselves. It follows that political coalitions will only support such standards if they expect the standards body to go on revising them until members receive approximately what was promised.

To see how this works, we return to our coffee example. With perfect knowledge of the industry's supply-and-demand curves, suppliers and retailers would know exactly what benefits and burdens they were voting for. But in fact, we saw in Chapter 2 that real standards seldom get this right the first time. For example, 4C and SFI seem to have generated too much supply while FSC produced too little. Figure 4.1 shows how communities might try to fix such mistakes. Suppose that the community's original standard was meant to reach an equilibrium at Point 1 but that the original demand estimate was too optimistic; then we would expect suppliers continue to drop out of the bargain until the actual equilibrium shifted to Point 2. At this point, governance bodies could reasonably ask whether the standard could be revised to more closely approximate the parties' original, democratically agreed target. Enacting a new and tougher standard that moves the equilibrium closer to Point 3 plausibly does this.

Trial-and-error methods may also be necessary because some parties have more information than others. Returning to the figure, anchor firms will almost always know more about demand for standards, while suppliers will know more about their willingness to supply compliant coffee at a given price. This gives both sides incentives to lie. For example, suppliers will almost always claim that the supply curve rises slowly, since this implies that anchor firms must offer more reward. Conversely, anchor firms will often overstate likely demand so that suppliers invest more than they otherwise would. Understanding these incentives to lie, both sides may decide that it is safer to take no action at all. Whether they attempt a standard will then depend on how nimbly the organization pursues additional cycles. From this standpoint, the existence of two dueling and semipermanent standards bodies provides a handy commitment mechanism for reassuring nervous coalition members that the required cycles of trial and error will actually occur.

in miniature. In the second stage, the resulting standards are introduced to the broader society and, provided they survive, can evolve further based on experience and feedback.

There are nevertheless important differences. For the **internal politics** case, every community member belongs to the same standards body. As a result, organizers must build winning coalitions that span a greater diversity of opinion. This necessarily increases the chances that organizers may fail to obtain agreement for any standard at all. But when they do manage to achieve internal agreement the resulting coalition is more likely to approximate opinion in the wider society. Thereafter, however, we have argued that there will be less pressure to evolve standards in light of experience.

The advantages of **external politics** provide the mirror image of these arguments. Because rival initiatives are (by definition) smaller and more partisan, the standards they produce are likely to be slanted and might not survive at all in the wider society. But if they do, we expect them to evolve rapidly from repeated contacts with the public to a faithfully Downsian image of mainstream opinion.

Finally, external competition offers insurance against the danger that a single, industry-wide body could be corrupted. If such an institution was captured, we would expect its competitors to reveal and criticize that fact.[44]

4.5.3. Benchmarking Performance

We have argued that private politics suffers from deep information constraints that make it less responsive to shadow electors. The question is, "compared to what?" Self-governance scholars point out that traditional governments are similarly hostage to "elite groups and leaders"[45] who can ignore "democratic responsiveness" for long periods.[46] More formally, agency politics is almost always dominated by "iron triangles" or, in modern usage, "subgovernments".* The power of these groups is limited only by elected officials' and/or the general public's ability to monitor and intervene. In practice, monitoring usually depends on citizen complaints,[47] while intervention depends on prioritizing scarce resources to those cases where intervention offers the most benefits. Intervention is also systematically less likely for issues like national security, whose constituencies tend to be geographically dispersed.[48]

Private politics plausibly improves matters along one or both dimensions. First, conventional politics is expensive and favors those who can afford lobbyists.† This suggests that, for the U.S. federal government at least, private channels are systematically more inclusive than public ones.‡ Second, monitoring by itself is insufficient: outsiders must also be able to

---

* See, e.g., Berry (1989) (subgovernments); Cater (1964) (same); Pulitzer and Grasty (1919) (iron triangles). The power of different subgovernments varies widely and is relatively disabled for topics like national security, whose partisans are geographically dispersed. McCubbins, Noll, and Weingast (1987).

† This is nicely illustrated by our artificial DNA case, where small firms seldom had the budget to attend agency meetings and sometimes learned about hearings after the fact. Private governance meetings were invariably more accessible and were often scheduled in parallel with industry trade fairs and other convenient venues.

‡ We expect incumbent interest groups to resist expanded participation where they can. Additionally, most agencies are culturally disposed to favor certain groups over others. This dynamic almost always favors incumbents. See, e.g., Moe (1998); Berry (1989); Harris and Miklis (1989); McCubbins et al. (1987); Ripley and Franklin (1976). These biases are further amplified by institutional inertia, for example when officials self-select into agencies they find congenial or adopt agency culture for strategic reasons like furthering career ambitions, pleasing coworkers, avoiding conflict, and reducing workloads. Wilson (1989). In keeping with our artificial-DNA example, we also expect U.S. government agencies to favor citizens over noncitizens.

credibly threaten intervention if the triangle's incumbents ignore them. In practice, this capacity differs widely: in some cases, at least, subgovernments are reportedly open to a wide range of viewpoints.* The question remains whether private bodies can do better. The fact that our forestry, fishing, and artificial-DNA examples produced standards where public channels deadlocked shows that they empowered new constituencies. Whether they also embraced more constituencies in toto is less clear. Our fisheries, coffee, artificial-DNA, and forestry examples all included instances where organizers deliberately excluded or disadvantaged some groups compared to others.

4.6. Private–Public Interactions

Private governance cannot be studied in isolation. Almost all real standards overlap with actual or potential government regulation. This raises the question of when and to what extent this coexistence improves policy outcomes for the combined system.

Coexistence and Crowding Out. For companies, developing private governance provides a hedge in case official regulation is inadequate, or late, or never arrives at all. But as time goes by, firms can rationally decide that the private channel is no longer worth the expense.[49] For shadow-of-hierarchy scenarios where the private standard is motivated by fear of public regulation, we expect firms to abandon private standards development when it becomes clear that government will probably enact strong regulations regardless. The case is different, however, where industry genuinely prefers some regulation and sees private standards as insurance against deadlocked or inadequate public rules. Here, industry will only abandon private initiatives when it sees that regulators are likely to produce an acceptable outcome.†

As usual, there is no guarantee that industry's incentives to abandon a private standard will produce the best social outcome. To the contrary, we have argued that private standards can strengthen public regulation by generating otherwise unavailable information on political feasibility, technology options, and affordability. This suggests that private and public governance initiatives should be encouraged to run in parallel for as long as possible.

---

* Where intervention is cheap, subgovernments are said to degenerate into "issue networks" that include a wide range of viewpoints. Berry (1989).

† In both cases, we expect firms' willingness to persist in developing private standards to increase as they approach completion so that less and less work remains to be done.

Public Intervention in Private Standards. Government usually has the last word. If it wants to, it can promote private standards with grants, government purchasing policy, and public endorsements. Our examples suggest that this has historically been done based on ad hoc judgments that a particular private standard will achieve government's own preexisting goals. However, officials could also adopt the more general principle that private democracy is likely to improve policy outcomes on net, or is ideologically desirable in its own right.

Conversely, government can criticize ("jawbone") standards it dislikes and even ban them within its borders. Whether such measures will influence worldwide standards is less clear. While large countries like the United States presumably exert considerable power, our fisheries example shows that the experience of small Scandinavian countries is mixed. More generally, we should expect that some standards will be backed by some governments and opposed by others. Finally, pressure is a two-way street. This suggests that conflicts may sometimes end with private industry coercing governments to change their official standards instead.* Such situations are especially likely in the developing world, where governments tend to be weak or else heavily dependent on rich-nation markets.

---

* In one recent example, Hollywood successfully threatened to stop making films in the State of Georgia unless the latter dropped so-called religious liberties legislation that many gay rights groups opposed. Schwartzel and McWhirter (2016).

PART III

ACADEMIC SCIENCE

5

Legacy: Academic Self-Governance in Modern Times

Part II constructed and explored a New Self-Governance paradigm in which private power and politics emerge from the purchasing power of large anchor firms. We now extend this baseline model to academic communities where resources are allocated by scientists themselves through peer review, committees, and similar nonmarket institutions. As before, we begin by summarizing the main historical examples; Chapter 6 uses this evidence to extend our power and politics theories to modern academic communities.

5.1. Asilomar (1975)

Prior to the late 1960s, the U.S. National Institutes of Health (NIH) had no formal system for flagging potentially hazardous proposals. Instead, biosafety practices were informal and varied from lab to lab.[1] This changed abruptly when three controversial research agendas brought researchers face to face with the new science of genetic engineering.

Paul Berg. Prominent Stanford biochemist Paul Berg knew that the simian SV40 virus made hamster and human cells cancerous. It was natural, then, to ask whether the virus could similarly transfer noncancerous genes.[2] In the summer of 1971, one of Berg's graduate students mentioned the idea at a Cold Spring Harbor workshop. The workshop leader, Robert Pollack, immediately saw that the idea might be dangerous.[3] What would happen, he asked, "if a human bacterium with a tumor virus inside escaped the lab and created a public health catastrophe?"[4] Pollack felt vulnerable because he had no tenure.[5] At the same time, the fact that he did not use SV40 in his own work meant that he could worry about Berg's proposed technology impartially, "without threatening my own career."[6]

Pollack invited Berg to give the workshop's closing lecture. When Berg declined,[7] Pollack escalated the issue by sending students home with a memo saying that "No one should be permitted to do the first, most messy experiments in secret and present us all with a reprehensible and/or dangerous fait accompli at a press conference."[8] Berg shot back that Pollack was "crazy," adding that his group would engineer precautions into *E. coli*.[9] But Pollack refused to back down, telling Berg that the work was "intrinsically dangerous" and that "We're all in over our heads."[10] In principle, Berg could have gone ahead regardless. Instead, he took the issue to fellow elites, including Nobel laureate James Watson (Cold Spring Harbor), David Baltimore (MIT),[11] Maxine Singer (NIH), and several NIH virologists who had worked with SV40.[12] To his surprise, most responses were critical.[13] Berg then postponed the experiment,[14] using the hiatus to raise the issue with still more scientists.[15] Europeans were noticeably more skeptical than Americans.[16]

Andrew Lewis. Meanwhile, NIH researcher Andrew Lewis discovered that the United States had inadvertently created a hybrid organism that mixed SV40 DNA with a virus that caused respiratory disease in children.[17] Lewis was young and unknown but had a secure position at the U.S. Public Health Service.[18] In early 1971, Carel Mulder (Cold Spring Harbor) asked Lewis for the hybrid so that he could see which parts of the SV40 genome coded for cancers. Lewis hesitated but promised to supply a sample by September. At first Lewis's motives were mostly competitive, i.e., making sure he could publish a discovery paper before Mulder wrote his own article. Soon, however, Lewis and his group began worrying about biosafety as well. When Lewis visited Cold Spring in August, Watson berated him for not sharing the virus and threatened to write letters of complaint to the NIH, *Science*, and the U.S. Congress.[19] Lewis responded by raising biosafety concerns, adding that recipients should acknowledge possible hazards and implement precautions.[20] He later complained that Watson had publicly chastised his young and vulnerable team.[21]

The argument continued at that year's Cold Spring Harbor Conference, provoking opinions ranging from "there's no problem here" to "the hybrids should never leave NIH."[22] Significantly, the publication and safety issues remained tangled. When Lewis sought advice from older scientists, they mostly disagreed about safety but saw no obligation to share the virus before publishing.[23]

Lewis eventually forwarded the promised sample to Mulder, adding that he should "consider" withholding it from other labs.[24] By then at least twenty labs had asked for the seed, some of which refused to honor Lewis's restrictions.[25] He therefore asked the National Institute of Allergy and Infectious Diseases (NIAID) to develop a formal policy statement. By

January 1973, most U.S. officials agreed that the problem required formal regulation.[26] Lewis therefore drafted a memorandum of understanding that required seed recipients to set up minimum containment standards and accept full moral and legal responsibility. NIAID approved the memorandum in November, although the legal basis was doubtful.[27] Despite the safety issue, many scientists believed that "the real intent [was] to allow NIH scientists to use these viruses while at the same time other scientists [were] denied use of them."[28]

Herbert Boyer and Stanley Cohen. The third and in many ways deepest biosafety issue involved University of California, San Francisco Professor Herbert Boyer and Stanford Professor Stanley Cohen, who had begun using pSC101 plasmids to transfer DNA between organisms. Unlike Berg's methods, the new technology was simple to use.[29] When Boyer announced the work at the Gordon Conference on Nucleic Acids in June 1973, audience members realized that gene splicing would soon be available to everyone.[30] Two British researchers immediately voiced concerns to Singer, the conference chair. While several biologists objected that controversy could invite restrictions,"[31] Singer responded that "if two level-headed researchers … could react so strongly" she would schedule a half-hour policy session to discuss it.[32] This turned out to be a pivotal moment: Singer later conceded that "the whole thing wouldn't have gone to first base" under a more conservative chair.[33]

The next day, Singer asked attendees for proposals. Suggestions included leaving the matter to individual conscience, calling for a National Academies panel, and writing to leading science journals. She then called for a show of hands without further discussion.[34] Seventy-eight of the roughly ninety-five people present voted to send a group letter to the National Academy of Sciences (NAS). Of these, forty-eight endorsed the still more public step of sending the same letter to *Science*.[35] Because many members had already gone home, Singer wrote a follow-up letter asking all 142 attendees whether they agreed. In the end, only twenty wrote back to oppose,[36] arguing that a letter could invite public attention and unnecessary restrictions.[37] Singer wryly recalled that younger researchers were noticeably more willing to raise social concerns – except where their own research was affected.[38]

Following the Gordon Conference, Cohen's lab received numerous requests for pSC101. At first he refused, sparking widespread resentment[39] and an angry quarrel with Berg.[40] By October, Cohen had relented and was sharing the pSC101 plasmid with any researcher who promised not to introduce tumor viruses or antibiotic resistance into bacteria.[41] Berg's lab was also deluged with requests for the plasmid,[42] including researchers proposing "horror experiments" like inserting tumor-producing herpes or diphtheria genes into *E. coli*.[43]

The Moratorium. *Science* printed the Gordon Conference letter on September 21, 1973.[44] Remarkably, the event went almost entirely unnoted by the public.[45] Despite this, the National Academies immediately asked Singer to suggest someone who could lead a study panel.[46] She nominated Berg, in part because he had agreed to postpone his SV40 work.[47] Berg refused sole responsibility, but agreed to talk with colleagues. By February, he had proposed a committee to discuss the new technology's risks and identify possible long-term actions. The politically mainstream group included Richard Roblin (Harvard), Baltimore, Herman Lewis (National Science Foundation), Watson, Sherman Weissman (Yale), Norton Zinder (Rockefeller), and Berg himself.[48] But when Cohen asked to join, Berg initially refused on the ground that the panel was limited to cancer specialists. Fearful that the absence of plasmid experts would "unwittingly penalize" his work, Cohen threatened to write his own letter. At this point Berg relented, saying that it would be "silly" to have two letters and promising to "work out the phraseology to please us all." He then invited Berg and three other Stanford professors to join the committee "to keep peace."[49]

The committee met on April 17.[50] By then, Cohen realized that he was losing control of the technology. This prompted some committee members to call for a moratorium on future experiments until the conference could meet. Berg recalls that the suggestion "came as a shock" and "seemed rather radical," but soon agreed.[51] Watson similarly objected that a ban was premature but then relented.[52]

The committee began drafting a letter to *Science*, *Nature*, and *Proceedings of the National Academy of Sciences* announcing a worldwide moratorium on experiments that used plasmids or viruses to transfer toxin-production capabilities or antibiotic resistance to new species. The letter also called on the NIH to form an advisory committee to develop safety procedures and convene an international meeting "of all involved scientists" in early 1975.[53]

Baltimore showed drafts of the letter to colleagues at the NAS, the European Molecular Biology Organization (EMBO), and Cold Spring Harbor without encountering opposition.[54] Shortly afterward, however, one of the signatories' students leaked a copy to *The New York Times*.[55] This forced a press conference,[56] along with modest coverage by the BBC, the Australian Broadcasting Corporation, and other media outlets.[57]

No one was sure that the moratorium would hold. While an informal phone survey suggested that most scientists "firmly endorsed" a temporary ban,[58] some observers were skeptical.[59] In the words of one journal editor,

"Caltech and Harvard will respect them, but those not in the elite will see no reason to hold off"; another biologist added that "Anyone who wants to will go ahead and do it."[60] However Baltimore disagreed predicting that even the "worst part" of the biology community "will be under a kind of moral pressure to go along with the majority."[61] Berg added that "Anybody who goes ahead willy-nilly will be under tremendous pressure to explain his actions."[62] Finally, *Science* was more tentative, speculating that the embargo would "[q]uite possibly hold until the [Asilomar II] conference" but its "real test" would come if biologists decided to extend it "indefinitely" after that.[63]

Beyond these very general arguments, relatively little was said about how to enforce the ban except in Britain, where the government ordered grant recipients to participate.[64] Baltimore told the press that editorial boards should refuse to publish experiments covered by the moratorium.[65] But while *Nature* expressed sympathy for the moratorium, it refused to say whether it would publish "reports from those who work on regardless."[66] In fact, the moratorium seems to have been "universally" observed.[67]

The committee's letter also touched off further controversies. University of Washington Professor Stanley Falkow had heard rumors of the Berg committee and obtained his federal grant officer's explicit approval to isolate the toxin gene in cholera.[68] Now he told Baltimore and Berg that they did not understand his work.[69] Berg responded by asking him to organize a session at what would become the definitive conference (Asilomar II) in February 1975.[70] Meanwhile, Alabama microbiologist Roy Curtiss had been trying to reengineer *E. coli* to fight tooth decay. The Berg letter persuaded him to stop work and send a sixteen-page letter asking more than a thousand scientists whether they saw a problem. Far from being angry, he leaned toward a still broader moratorium.[71] But this position was controversial. British doctor Ephraim Anderson protested the moratorium letter in the next issue of *Nature*,[72] while Lederberg told *Science* that the letter could invite "impediments" and remove "the ultimate decision from the hands of scientists."[73]

Asilomar I. Berg nominated Pollack to organize a conference on the hazards of tumor viruses.[74] This led to a meeting of about one hundred scientists at California's Asilomar Conference Grounds during January 22–24, 1973.[75] All but one or two attendees were American and the press was not invited.[76]

For the most part, the proceedings focused on narrowly technical discussions to identify biology risks. However, some attendees voiced concerns that the issue would unnecessarily frighten the public and that funders would force scientists to implement new safety measures from existing budgets.[77] Other attendees argued that all biology research entails risks, and that the rewards of genetic engineering would be justified if even "five or ten" people died. Adopting precautions would also delay research and increase costs. Finally, Watson argued that government officials, not scientists, should decide the issue. He added that the "NIH has to face up to paying for the costs of safety or declaring the viruses we work with as not dangerous."[78] But he also said that academic scientists had to behave responsibly, and that he himself would go to court to get an injunction if workers in his labs did work he thought was dangerous.[79]

Organizing Asilomar II. Berg, Baltimore, Singer, an NIH observer, and prominent European biologists Sidney Brenner (Cambridge) and Niels Jerne (EMBO) met on September 10, 1974 to organize what would become the definitive Asilomar II conference.[80] One of the group's first decisions was to focus narrowly on the biosafety issues that bench scientists needed to address to do their work. This deliberately avoided broader social and ethical issues that might have stoked a public controversy.[81] Attendance was set at 150 to ensure a manageable outcome, of which roughly one-third came from abroad.[82] The NAS agreed to host the meeting with NIH support.[83]

The leaders also named chairs for three groups tasked with investigating specific technologies and reporting back recommendations.[84] In addition to basic science credentials, candidates had to "carry considerable weight with ... colleagues" in case they decided that some experiments should be banned.[85] Recruitment then proceeded chain-letter fashion, with chairs picking further members, who in turn suggested additional invitees and sometimes deletions: "The process of selection depended heavily upon informal networks of a few key biologists."[86] These included researchers whose plans would be most affected by regulation[87] along with dissidents like Curtiss and Falkow who had pushed back against the Berg letter.[88]

The organizers also invited nonscientists. These included a handful of industry and law speakers[89] and a dozen or so reporters. Berg later credited press coverage with "encourage[ing] responsible discussion that led to a consensus that most researchers supported."[90] One activist group refused an invitation as tokenism[91] and demanded that the moratorium remain in effect until the public could participate.[92] The organizers ignored it.[93]

Asilomar II. The conference took place between February 24–27, 1975.[94] Berg later recalled that the meetings were "Confusing, because some of the basic questions could only be dealt with in great disorder, or not confronted at all."[95] Berg was also "struck by how often scientists willingly acknowledged risks in other's experiments but not their own."[96]

The Organizing Committee wanted to keep oversight as much as possible within the scientific community.[97] However this meant that Berg had to find a way of "feeling out the consensus of the participants." At the same time,

Baltimore worried that a "no confidence" vote would be disastrous.[98] He therefore opened the meeting by warning that members should not expect a vote. Instead, the Organizing Committee would present "consensus" recommendations on the final morning.[99] The Committee also worried about public expectations. As Berg later remarked, "[Y]ou could see that people recognized that if we came out with nothing it would be a disaster."[100] Failure to extend the moratorium would be particularly embarrassing.[101]

The first three days were devoted to lectures and working sessions[102] with "heated discussions carried on during the breaks, at meal times … and well into the small hours."[103] Most of the formal input revolved around the three working groups:

> The **Plasmid Group** delineated six categories of experiment, each requiring different containment facilities with one category banned altogether.[104] The report touched off a fierce floor debate over whether the conference should produce any recommendations at all.* Hearing this, Brenner objected that individual institutions should not be allowed to self-govern without government supervision.[105] He threatened to resign and "repeatedly warned" that "doing nothing" would invite public condemnation and government interference.[106]
>
> The **Animal Virus Group** initially argued that there were no new risks and that existing rules were sufficient. However, Lewis wrote a minority statement arguing that the tempo of research should be deliberately slowed until more was known and new, inherently safe vectors became available.[107] This sparked more expressions of dissatisfaction from the floor,[108] which were further reinforced when lawyer Roger Dworkin told the conference that failure to act would invite Occupational Safety and Health Administration fines and "multimillion dollar" negligence claims.† The group ultimately reversed itself, amending its report to endorse physical and biological containment.[109]

---

* Anderson objected that the group's members did not know much about handling organisms (Rogers (1977) at p. 63), adding that everyone should go home, brood, and offer suggestions in writing. But when Berg asked whether the conference should adjourn without producing a statement, Anderson conceded that he didn't know. Berg pressed his advantage by arguing that silence would mean "telling the government to do it for us." (Id. at p. 76.) Lederberg argued that the proposal was too precise and would invite legislators to make it permanent. Watson and Lederberg, the only two Nobelists present, favored repealing the moratorium before it hardened into a permanent ban. Frederickson (1991) at p. 278 (Watson); Lear (1978) at p. 131 (Lederberg). Singer pushed back hard by asking Watson what had changed since he had agreed to the moratorium in April. Rogers (1977) at p. 64.

† Rogers (1977) at pp. 78–81; Lear (1978) at pp. 140–2. Dworkin added a more positive incentive by explaining that courts often defer to private standards. Id.

The **Eukaryotic Recombinant DNA** working group recommended three levels of containment depending on perceived risk. This framework was later adopted in the conference's final recommendations.

The audience reacted suspiciously. Watson whispered to his neighbor, "These people have made up guidelines that don't apply to their own experiments."[110] Conversely, participants whose work was affected were quick to protest the classifications.[111] Berg later recalled being "struck by how often scientists willingly acknowledged risks in other's experiments but not their own."* In the words of one anonymous participant, "The consensus here is that people want guidelines and containment so they can go and do their experiments, but no one will come out and say it – they're all chicken."†

Strikingly, it was almost impossible to know what average members thought. Instead, most of the input to the final conference statement came from about thirty people, including the organizers and working group members.[112] The rest were largely "passive or reactive,"[113] hoping that prominent and senior people would speak for them. "Stand up and say it," one biologist told Watson. "You can say it; I can't."[114] Berg later realized that "the Lederbergs and the Watsons and a few others were doing all the talking" but did not actually reflect "what everybody wanted and felt."[115] Berg was "worried because I could not pick out a common thread. I could not see the consensus."[116] He later said that he was surprised by the eventual vote, while Singer confessed that she had been a "rotten politician."[117]

The Final Session. The science sessions were now complete. The committee decided to submit a statement even if it was voted down,[118] completing the draft at 4.30 AM on the conference's final day. The document had "considerably more bite" than any of the working-group schemes. It proposed doing most experiments in "moderate" or "high-risk" labs with special equipment, although no experiments would be banned entirely. Most experiments also required suitably crippled host bacteria, which would not be available for months. This artful compromise simultaneously extended the moratorium and ensured that it would end within a predictably short time frame.[119] Participants were allowed roughly thirty minutes to read the document.[120]

---

* Berg (2008). Berg later added that "everybody [wanted] to draw a circle around their work and stamp it as pure and unadulterated – it's what you're doing that is nasty and needs to be proscribed." Rogers (1977) at pp. 82–3.
† Wade (1977) at pp. 47–8. Even if members agreed, the new precautions were bound to be burdensome. Berg's recent lab upgrade, which only satisfied some of the proposed categories, had cost five figures. Rogers (1977) at p. 61. Similarly, Watson had argued at Asilomar I that upgrading a moderate-size lab would cost about $125,000 and "much more" for "a major lab." Hellman, Oxman, and Pollack (1973) at p. 351.

Berg opened the session by saying that the recommendations were provisional.[121] Still afraid that the audience would vote down the statement,[122] he immediately added that the Committee would be solely responsible for the text, albeit based on "having listened to the debate and the discussions"[123] and its sense of consensus.[124] He added that none of this required a vote.[125] At this point, Cohen rose to demand a show of hands.[126] David Botstein (MIT) mediated the crisis by proposing a test vote, reassuringly adding that up to "eighty percent" would probably support the document. Once the press had been ushered out of the room, Brenner called for a show of hands on the first paragraph.[127] To Berg's surprise, the yeas were "virtually unanimous" and no one opposed. However, one member protested that the principle that some experiments should be banned entirely had been left out.[128] The point would be repeatedly raised from the floor for the rest of the morning until the final vote, when Anderson announced that he considered it "imperative." All further opposition collapsed at this point: participants instructed the Committee to insert the provision with just five dissenting votes.*

The second part of the statement endorsed a framework in which containment strategies depended on risk. However the method of estimating risk was left for national bodies and hedged with caveats[129] This vote also passed with just token opposition.[130]

The next section described low-, medium-, and high-containment facilities based on existing National Cancer Institute safety standards. This time "[t]here was little opposition ... but many abstentions."[131] Despite a sharp exchange with Lederberg, Berg announced that the committee would report "overwhelming consensus."†

---

* Baltimore initially conceded that that the committee's decision not to ask for a ban was "certainly reversible," suggesting that the document should "acknowledge that the scientific community is split on this issue." Rogers (1977) at pp. 90–1. Two researchers (apparently Anderson and NIAID's Robert Martin) raised the issue again later in the session by demanding a test vote. Rogers (1977) at p. 93; Lear (1978) at p. 142. At this point, Berg conceded without a vote "that there is a class of experiments that should not be done at all." Rogers (1977) at p. 93; cf. Frederickson (1991) at p. 282. However, other participants continued to argue that banning experiments was a dangerous precedent. Krimsky (1982) at p. 146. The final showdown between Anderson and Berg is described in Lear (1978) at pp. 144–5.

† When Berg announced "It's clear, that the sentiment is overwhelmingly in favor," Lederberg immediately objected "I don't think that's clear at all, Paul." He then added that "very complex issues are being railroaded through" – Berg interjected "nope" at this point – "and the assumption is being made that there is a consensus on issues that have simply not been ventilated." Berg used his position as chair to have the last word, arguing, "But I have to take a show of hands as an expression of the sentiment ...; I saw two or three noes, and I didn't count the number of ayes. So I have to call that overwhelming." Rogers (1977) at p. 94.

The fourth part of the statement, calling for the development of safer vectors, lab training, and education, was "almost totally uncontroversial."[132] The lunch bell rang while the discussion was still under way, prompting Berg to move to the final section, listing further research questions that would advance safety. No one had anything to add.[133]

By then the second lunch bell had rung and Berg asked for "a show of hands" for what he called "a fair consensus, without unanimity." Cohen objected, "Paul, I can't vote on that until I see the wording of what we're supporting." Berg ignored him.[134] Lederberg, Watson, Cohen, and Philippe Kourilsky (Pasteur Institute) voted nay.[135] Anderson criticized the meeting as "a bunch of people with no experience in the handling of pathogens" but nevertheless voted yea because "[o]ne had to leave the matter open at that moment."[136] "Okay," said Berg. "Again I would say that there is substantial agreement here on that point."[137]

Aftermath. The Organizing Committee repeatedly promised to consider additional written submissions after the conference.[138] Watson publicly called Asilomar a "waste of time," adding that scientists had "pretended to act responsibly but actually were irresponsible in approving recommendations that did not adversely affect the work of anyone present."[139] He added that putting experiments in high-risk facilities meant "discouraging them altogether."[140] The Plasmid Panel similarly dissented, declaring itself "very angry" that the final statement had ignored them and adding that virologist members had protected their own interests by putting "the most stringent prohibitions … on the people who worked with microorganisms." The Organizing Committee refused to publish the panel's report as an annex to the final statement.[141]

NIH. Berg delivered the Organizing Committee's final report to the National Academies on April 29, 1975.[142] NAS approved the final Statement in May and published it in June.[143] Berg, fearful that the moratorium would unravel, urged the NIH to convene its Recombinant DNA Advisory Committee (RAC) quickly.[144]

That October, the RAC met to elaborate the Asilomar consensus into specific safety procedures.[145] Strangely, the first draft tried to reconsider the principle that some experiments should be banned entirely. This met with fierce criticism from scientists and was withdrawn.[146] In the words of one observer, "Asilomar had indeed identified the center of gravity of safety precautions. When a shift from that center of gravity was attempted … the scientific forces were sufficiently mobilized to bring the guidelines back to its 'equilibrium' position."[147] A second, more stringent draft was finalized in July 1976.[148] Thereafter, scientists who persisted in criticizing the RAC guidelines reportedly suffered "backlash [that] ranged from overt character assassination to a more subtle form of professional ostracism."[149]

5.2. Mutations Science: Building a Community-Wide Database (1999–2001)

Biologists spent the 1980s and '90s learning to transcribe the blueprint ("sequences") recorded in every DNA molecule.[150] But the result was like trying to read an unknown language: scientists still needed to find out what processes each sequence controlled inside the body. One of the best clues was hospital data from patients whose DNA had been damaged by specific mutations. To the extent that this information had been collected at all, most was housed in databases run by individual biologists as a labor of love. But while the biology content was excellent, the system suffered from incompatible definitions ("nomenclatures") and primitive, homemade computing architectures. This made it hard to compare or combine information across databases. Finally, there were not nearly enough of them. By the early 2000s, there were roughly one hundred databases covering 270 genes. This was less than 1 percent of the disease- and cancer-causing mutations thought to exist. While some individual groups had tried to combine the information into larger datasets, community members universally agreed that the results barely scratched the surface of what the databases could offer if they were unified.

Self-Governance. In 1994, the March of Dimes* (MoD) funded a community organization called the Mutations Database Initiative (MDI) to construct a single, worldwide database to fix these problems. The catch was that MoD would not support the project forever: at some point, MDI would have to become self-sustaining. For the first five years, the project concentrated on getting members to agree on a standard nomenclature, entry form, and model content for mutations data. These first steps already produced passionate debate. This was understandable, since nomenclature choices shape which questions can be asked and, to that extent, necessarily favor some researchers compared to others. That said, academics often argue about nomenclature. This made the disputes manageable and familiar.

Building an actual facility would be harder. From its founding, MDI had repeatedly sought grant support from governments and medical charities across the United States and Europe. But those efforts had failed. By the late

---

* The March of Dimes is a leading U.S. health charity.

1990s, MDI's leaders had concluded that further applications were hopeless and that the only remaining choice was to find a commercial partner. This meant that the community would have to decide what sorts of business arrangements were acceptable. MDI's private politics were about to become much more complicated and controversial.

The Players. By 2000, MDI included more than six hundred mutations scientists from thirty-three countries. Nearly two hundred followed the organization's activities on a regular basis, of whom fifty or so typically attended twice-annual meetings. Members could conveniently be grouped into three distinct populations:

Organizers. The driving force behind MDI came from the late Richard G. H. Cotton[151] with advice and public support from prominent geneticist Charles Scriver.[152] Cotton led all of MDI's public meetings, crucially deciding the content and timing of votes. MoD executive Michael Katz attended meetings and was included in correspondence, but made few comments beyond repeated statements that his foundation could not fund MDI forever.

Large Players. MDI's Large Players consisted of five large, well-funded academic groups that specialized in constructing advanced biology databases and software. These included the Human Gene Mutation Database at the Institute of Medical Genetics in Cardiff, Wales (Cardiff), HGBASE at Karolinska Institute in Stockholm (Karolinska), the European Bioinformatics Institute (EBI) at Cambridge University, Mitomap at Emory University (Emory), and the Genome Database at the Hospital for Sick Children in Toronto (Toronto). A sixth organization, Johns Hopkins's Online Mendelian Inheritance in Man project, was in contact with the MDI but never actively participated. Compared to the average MDI member, Large Players were extraordinarily well funded and computationally sophisticated.

LSDBs. Individual MDI members ran Locus Specific Databases (LSDBs) that studied specific mutations in depth, usually without grant support. Members hoped that a single, unified database would make this work more visible and useful. Unlike the Large Players, LSDB operators had little interest in computing. To the contrary, they wanted a central site to take over these chores so that they could spend more time doing biology.

MDI's leaders knew that they needed to find out what the community wanted and build consensus for an eventual deal. They used MDI's twice-yearly meetings as an occasion to report progress and vote on next steps.

San Francisco. In October 1999, Cotton convened roughly a dozen Large Players to discuss the possibility of a commercially supported mutations database. The group included Cotton, the author, Katz, various LSDBs, and commercial data provider Incyte Corp. Most agreed that the project was economically and technically feasible and deserved further study. Only one Large Player (Emory) was ambivalent, expressing doubts that the proposed facility could be built. The next day, eighty MDI members met for their regular semiannual meeting. NIH representative Lisa Brooks confirmed that her agency "had too small a budget to fund new databases." Most of the meeting was spent discussing commercial support.

Then and now, the idea of academic communities partnering with corporations was unfamiliar. This led to endless debates about whether the community could obtain a better offer. Since members had no experience selling biology data, the arguments were formless. Partnering with commercial firms also raised ideological issues. While most members ignored these arguments, a few insisted that any agreement respect "open-source" principles of free access. Cotton arranged for the meeting to debate both issues at length. In the end, the most convincing argument came from an Incyte employee who suggested that his company would probably pay a substantial fee for better mutations data. This gave the commercial option immediate credibility.

Cotton ended the meeting with a show of hands, confirming that MDI should pursue commercial support. However, he left open the possibility that traditional grant support might also be possible. The vote preserved this ambiguity by authorizing MDI's leaders to "work as a group to induce funding" so that the organization's activities could be "raised to a higher level." The "final plan" would be voted on when the community met six months hence in Vancouver.

Cotton then convened a twenty-one member working group including MDI's leadership, the author, Large Players at EBI and Cardiff, a MoD executive, and several LSDBs. The group spent the next few months drafting detailed "Specifications" designed to show commercial partners that the community could negotiate on normal business terms. This notably included acknowledging that the proposed depository might need to offer premium services to paying customers.* Despite this now plainly commercial focus, Cotton continued to preserve a measure of ambiguity by

---

* Meanwhile, academic and nonprofit users would receive basic access to data and a large number of powerful "Core Tools" free of charge.

specifying that the project would seek "traditional public and/or private sector grants ... to the maximum extent possible."

The sole dissenter was Large Player Cardiff, which complained that the draft should have emphasized its role as "THE [sic] potential central mutation data repository" and demanded that Cardiff – not MDI – retain all profits and make all marketing decisions including the right to sell "exclusive rights ... to any interested party."* Cardiff resigned from MDI's "stakeholders' list" in February. Several months later, it sold its rights to Incyte. The NIH remarked that the Specifications looked "promising" but complained that other private/public alliances had put their data in the public domain.

By February, eighty-six community members had reviewed and commented on the Specifications. Despite occasional qualifications, the comments were unanimously favorable. Probably the most important came from biotech executive Maureen Kelly, who argued that it would be simpler to launch the Depository as a "consortium" with industry. MDI's leaders agreed, advised members by e-mail, and continued to discuss the project.

Vancouver. MDI held its next scheduled meeting on April 9 in Vancouver. Thirty members attended the six-hour session, including Large Players Emory, EBI, GDB, and MutationView. Roughly half the session was spent in a chaotic debate over whether MDI should take the "next step" of negotiating a deal with industry. Without actually opposing the project, Emory was by far the most vocal. First, it reprised its earlier argument that unifying existing LSDBs might not be technically possible. Second, it argued that industry would eventually offer a better deal, possibly including cash grants to individual LSDBs. Despite this, the overwhelming sentiment from the floor was to go ahead. As one participant memorably remarked, "What choice do we have? Let's try it and see." Emory was still on the floor raising possible issues when the cocktail hour arrived. With the first members already leaving the room, Cotton called for a vote. The specifications passed by acclamation with Emory neither dissenting nor agreeing. Members also passed a resolution authorizing Scriver, Cotton, and the author to negotiate a draft agreement with industry in time for MDI's next meeting in Philadelphia.

---

* Cardiff also demanded that MDI give it "due recognition as the core of the future MDI facility" and that "MDI should start to recognize [Cardiff] as the de facto, central mutation database around which the planning of any future expansion/improvement ... should take place."

Despite the chaos, the vote made the Depository project dramatically more credible. Moments after the meeting, EBI's representative asked if his group could submit a proposal to build large parts of the database.* That same evening, GDB offered to build a prototype, pilot-scale database to test Emory's technical objections. On May 5, MDI's leaders formally invited the remaining Large Players to submit proposals. This led to a joint $2.3 million proposal by which construction and operation costs would be split among EBI ($1.6 million), GDB ($300,000), and Karolinska ($400,000). In the meantime, MDI's leaders asked rank-and-file members for written pledges confirming that they would donate data, editing, and/or administrative support to project. In the end, 123 members, including sixty-six LSDBs, agreed. This included practically all of MDI's active members, suggesting near-unanimous support for a commercial partnership.†

Negotiations with Incyte. MDI submitted its proposal to Incyte in mid-July. Detailed meetings followed, with Incyte delivering a proposed Memorandum of Understanding the night before the Philadelphia meeting. Under the Memorandum, Incyte committed to funding approximately $1 million in costs directly and find additional corporate partners (several of which had already expressed interest) to pay the balance. In return, MDI would promise to (a) construct the Depository so that it would be compatible with Incyte's Genomic Knowledge Platform (GKP) middleware standard, (b) write new GKP tools for the project, and (c) make Incyte the only commercial website to host the Depository. Otherwise, the Depository would be freely available to both academic and commercial users. The agreement would last three years.

MDI's leaders reported details of the emerging deal with Incyte to Emory, EBI, and the NIH in late August. The NIH's Francis Collins declared himself "pleased to hear the reassurances about broad access," but kept a careful silence about whether he thought the agreement appropriate. Meanwhile, EBI explored a separate deal with Pharmacia while Emory negotiated with a private/public development agency called the Georgia Research Alliance. Neither discussion had produced a concrete proposal by the time MDI met in October. The organizers refused Emory's last-minute request to put off the planned Philadelphia vote until some later date when Emory had "an

---

* There may have been a special reason for this. The European Union had recently slashed EBI's budget by 45 percent, demanding among other things that EBI find more commercial support. Balter (1999, 2001); Butler (1999).
† One LSDB declined to pledge, citing concerns over possible intellectual property rights.

alternative proposal" to offer. Meanwhile, EBI became increasingly critical of any deal with industry, telling Incyte in September that "the community" would insist on open-source principles. On September 28, EBI told MDI's leadership that Incyte's offer, was only "a first negotiating position" and should be rewritten as a smaller deal that would provide "seed money" for EBI alone.

Philadelphia. Seventy-two members attended MDI's Philadelphia meeting on October 3. The first part of the meeting was open. Incyte described its offer while EBI and Emory presented their own (unfunded) plans. This was followed by a closed, three-hour meeting where Emory and EBI violently denounced the project. Emory stressed that MDI could only act unanimously since a majority vote would ask dissenters "to go against [their] conscience." It also claimed that it could negotiate a better deal worth "tens of millions," possibly including endowed chairs for some non-Emory members. In any case, the community could always go back to Incyte if Emory failed. EBI's objections were more ideological, claiming that "open-source" principles required that Incyte let competitors host the database on their websites.

Nearly all members had come to Philadelphia pledging to support the project. But this did not include an open breach with Emory and EBI. Instead, members temporized, arguing from the floor that Incyte's proposal "would always be there" and that saying "yes" to a new type of private/public transaction would invite criticism. The turning point came when Cotton asked the NIH's Brooks to speak. Without once mentioning Incyte's proposal, she announced that the NIH might fund a central database after all. In the end, no vote was held. In the words of one observer, "I think the meetings in Philadelphia degenerated, in the absence of strong management leadership, into a situation where everyone felt defensive and insecure, and I doubt if many of those involved were proud of much of what happened." Neither EBI nor Emory ever put together a funding proposal and the NIH continued to turn down grant applications.

As one leading player remarked, "I would take the opinion that it was the vocal minority that let us down and not Incyte. This is a perfect example of how a community that whines and does not have a mechanism to make a quick decision can lose out."

5.3. Synthetic Biology: Academic Initiatives (2006)

The rise of companies able to manufacture cheap, gene-length DNA created a new, self-identified community of "synthetic biologists" around 2000.[153]

They claimed that artificial DNA would let them conduct experiments that would not otherwise be possible, including exotic, machine-like organisms to make jet fuel or hunt down cancer cells inside the body. The press loved the story. This was good for funding but also increased the chances of bad publicity and scandal.

The risks were obvious. Indeed, at least one lab had realized that it could synthesize smallpox DNA as early as 1999.[154] These concerns became more urgent after 9/11. By 2002–3, government officials[155] and commercial firms[156] had begun asking synthetic biologists to think about the risks and possible solutions. Prominent biology professor Eckhard Wimmer went even further, using artificial DNA to make the world's first artificial polio virus.[157] Though widely criticized,[158] he defended his work as a necessary "wake-up call" for scientists and society alike.[159]

These disparate threads received official recognition in 2004. That year, a National Academies panel called on scientists to "take responsibility" for "preventing potential misuses of their work,[160] the British Royal Society called for self-governance,[161] and the community-wide First International Conference on Synthetic Biology (SB1.0) held a "moderated discussion" of risks.[162] These were further reinforced the following year when the Sloan Foundation funded a $570,000, fifteen-month program to develop "options" – though no formal recommendations – for policy.[163] This made it natural for members to predict that the SB2.0, conference scheduled for May 2006 would take concrete actions on behalf of the community.* These calls had special resonance in a field that claimed descent equally from biology's Asilomar-style town halls and the electronics industry's philosopher-kings.

But the opportunity for private governance was fleeting. If SB2.0 failed to act it would reduce expectations for every SB conference thereafter. At the same time, it was plainly impossible for two hundred synthetic biologists to reach consensus in the space of a three-hour meeting. Working with conference co-organizer Jay Keasling, the author's UC Berkeley project polled roughly two-dozen synthetic biologists to identify ideas that could

---

* Prominent community member George Church predicted that
>A code of ethics and standards should emerge for biological engineering as it has done for other engineering disciplines. The community recognizes this need, but discussions are fragmentary. The next international meeting on synthetic biology (in May 2006 at the University of California, Berkeley) should make significant progress in that direction.

Church (2005); see also Carlson (2005); Service (2006) ("The Synthetic Biology 1.0 meeting held 2 years ago brought these issues to the forefront").

be implemented by a community-wide vote. The survey identified six initiatives that had been widely discussed and enjoyed broad support:

- *Mandatory screening.* The community could demand that artificial-gene companies screen incoming orders and boycott firms that failed to do so.*
- *Improved technology.* The community could construct better threat lists and advanced technology screening software.
- *Establishing norms: obtaining advice.* The community could remind members to obtain advice from independent experts before embarking on controversial experiments.
- *Establishing norms: reporting dangerous behavior.* The community could remind members that they had "an ethical obligation to investigate and, if necessary, report" dangerous behavior to authorities.
- *Clearinghouses.* The community could establish a "confidential clearinghouse[] to collect, analyze, and disseminate" information about safety and security risks.
- *Lobbying government.* The community could ask agencies to fund new safety technologies like inserting attribution data into DNA ("watermarking"), and engineering "inherently safe" host organisms that could not survive outside the laboratory.[164]

In April 2006, community members debated these proposals in two town hall meetings webcast from U.C. Berkeley and MIT. Roughly three-dozen scientists – one-fifth of the community – participated. Both meetings held votes calling for SB2.0 attendees to debate and vote on the first four resolutions that May.[165]

Some weeks later, however, conference co-organizer Drew Endy suddenly demanded that the vote be cancelled. Keasling and several nonbiologist policy professionals promptly called a meeting where they discussed and eventually agreed. While Endy never explained his reasons, other participants reported that they had been moved by the absence of a "constitution" authorizing voting and the possibility that a vote would invite public attention, government scrutiny, and/or "split the community."[166] Strangely, the turnabout was never publicly announced.†

---

* Drew Endy pointed out as early as 2003 that academics could refuse to do business with gene-synthesis houses unless they could "assure us that [they were] not synthesizing known threat agents." Maurer (2011(a)) at p. 1426. Endy later adopted the policy for his own lab. See Anon. (2005) at p. 5 (reporting that lab only did business with firms "that operate transparent procedures for screening gene-synthesis orders for potential bioweapons").

† The turnabout led to a confused sequel in which thirty-five activist groups redundantly called on synthetic biologists to cancel the vote. They have taken credit for the cancellation

Press coverage of the reversal was mildly unfavorable,[167] while scholars saw it as a "failed attempt."[168] Meanwhile, Endy tried to soften the disappointment by offering an online declaration that generally tracked the original conference proposals.[169] Ironically, the declaration was itself never finalized. Despite this, it probably encouraged work on improved screening technologies even though, in the end, no software was produced.[170] Later that summer, the Federation of American Scientists hosted a workshop that began designing new software. By fall, industry had formed a collaboration called the International Consortium for Polynucleotide Synthesis as an umbrella organization for moving the agenda forward.[171]

No SB conferences have attempted self-governance since. Some members have suggested that it would make sense to create an entirely new professional society authorized to hold votes and take action.[172]

5.4. Biology Journals: From 9/11 to Avian Influenza (2001–2011)

The most recent efforts to organize self-governance in biology have revolved around "experiments of concern" that might let terrorists and rogue nations construct cheaper and more capable biological weapons. The initial impetus came from government. Shortly after 9/11, the Department of Homeland Security and the White House announced that they were developing regulations to control the "discussion and publication" of nonclassified research that nevertheless posed a security threat.[173] Alarmed, the American Society for Microbiology asked the NAS to explore voluntary alternatives.[174] Echoing our nuclear-physics example, the focus soon converged on the possibility of organizing an editors' conspiracy. This culminated in a high-profile August 2002 meeting of academic biologists, security professionals, and journal editors.[175] The group debated a proposal to restrict publication for six types of research that could be used to make practical weapons,[176] but failed to reach agreement in the face of opposition from universities, journals, and scientists. The deadlocked group then threw the question back to the academic community by calling on "bioscientists and their professional organizations [to] take the lead in informing security experts how best to meet the threats of biological warfare and terrorism" without unduly harming research.[177]

Government officials renewed the pressure at a second National Academies meeting in January when they called on the scientific community

---

ever since. ETC Group (2006). Some authors have mistakenly assumed that synthetic biologists dropped their plans for self-regulation because of the e-mail. See, e.g., de Vriend (2006) at p. 65.

and journal editors, to "devise a better process" for handling unclassified research. This time, the statement was accompanied by an explicit threat that scientists needed to "come up with a process before the public demands the government do it for them."[178] Journal editors responded "that they hoped to release a joint statement shortly."[179] The resulting statement was published on February 21, 2003.[180] The signatories included sixteen past or present editors of leading science journals, three U.S. government agency heads, two security intellectuals, five academic biologists, and a free-speech activist.[181] But while the group conceded that information should sometimes be suppressed, it left the decision to individual editors.[182] This implied an obvious weakest-link dynamic in which the community could only suppress papers if every single editor agreed. At the time, it was still possible to think that editors would close the loophole by developing collective institutions for deciding when papers should be suppressed, for example by convening advisory panels of security experts and academic scientists. Indeed, some biologists had already called for such rules before 9/11.[183] But this was not done.

The crisis came in 2005 when *Science* Editor-in-Chief Donald Kennedy published an editorial defending his journal's decision to publish a synthetic-biology paper whose authors had resurrected the 1918 influenza virus. The first part of Kennedy's defense – that *Science* had consulted with the heads of the Centers for Disease Control and Prevention, the NIAID, and the NIH's Office of Biotechnology Activities – was unexceptional. The problem, Kennedy went on to explain, was that the NIH had then asked *Science* to contact the NSABB as well. Because *Science* had complied, one might have expected Kennedy to congratulate himself. Instead, he loudly claimed that his "convictions" would have "absolutely" led him to publish the paper even if NSABB had wanted it suppressed.[184] In effect, Kennedy was publicly announcing that he would publish what he liked – and daring critics to criticize him. Three years after 9/11, no one cared enough to press the issue. As Michael Selgelid remarked, the editors' extended flirtation with self-censorship had ended in an "unacceptable" result.[185]

Strangely, this was not the end of the story. In late 2011, the editors of *Science* and *Nature* delayed publishing new bird flu experiments after a government panel warned that the information could pose a security threat.[186] The crisis ended several months later when the panel withdrew its objection on the oddly circular ground that the U.S. government lacked any formal authority to suppress publication.[187] No one can say whether *Science* and *Nature* would have agreed to a permanent moratorium: Even so, the possibility of editors' conspiracies seemed to be making a comeback.

6

Academic Self-Governance: Power and Politics

This chapter extends our self-governance discussion to academic science. As in our commercial examples, we start by identifying a mechanism that lets sufficiently large majorities punish dissent. On the one hand, this makes private power and collective action possible. On the other, it supplies the private "constitution" against which all further politics play out.

Needless to say, academic self-governance presents significant differences compared to our baseline commercial case. The most obvious is that academic scientists do not interact through markets. Following the economics literature, we argue that they nevertheless exchange significant value through nonmarket trust and reputation networks. Conflicts that "split the community" damage these networks, reducing the rewards that members can reasonably expect from possible future trades. Crucially, these losses fall disproportionately on dissenters.

We proceed as follows. Section 6.1 reviews what is known about the idiosyncratic "Republic of Science" institutions that have grown up around Western academic research since the nineteenth century. It also extracts stylized facts from the examples presented in Chapters 1 and 5. Section 6.2 reviews the economic theory of trust and reputation networks. Section 6.3 extends these ideas to include academic communities whose members take turns allocating resources to one another based on agreed principles ranging from "scientific merit" to "old boy" cronyism. We focus in particular on how votes that threaten to "split the community" impact dissenters, providing the coercive basis for private power and collective action. Section 6.4 describes how the resulting politics plays out among organizers, players, and rank-and-file members. Section 6.5 asks whether recent claims that injecting nonscientists into private academic politics would make the Republic of Science more responsive to society.

6.1. Key Facts

This section identifies the key facts that any successful theory of academic power and politics should respect and explain. Section 6.1.1 sets the stage by introducing the modern Republic of Science institutions that let academics decide how research funds are allocated within their communities. Section 6.1.2 and Box 6.1 collect some of the striking commonalities that private academic politics displays across our case studies (Chapters 1 and 5).

6.1.1. The Republic of Science

Modern academic science depends on massive financial support by governments. However, officials typically know very little about which scientists and proposals are worth supporting. By comparison, individual scientists have far more information about the science potential of their own work and, to some extent, that of their colleagues. The challenge for regulators is to obtain impartial science judgments from these self-interested participants.

Origins. Nineteenth-century scholars conceptualized science as an autonomous force outside deliberate social control. This view reached its definitive statement in the 1940s, when chemist Michael Polanyi famously argued that letting researchers "freely make their own choice of problems" would create a self-governing Republic of Science. The community would then be "guided as by 'an invisible hand' toward the joint discovery of Nature."[1] For Polyanyi, the policy implications were obvious: non-scientists' attempts to prioritize research were impossible and nonsensical. "You can kill or mutilate the advance of science," he wrote, but "you cannot shape it."[2]

This argument was straightforward so long as science funding consisted mostly of paying researcher salaries. But that changed at the start of the twentieth century when the Carnegie and Rockefeller foundations began funding previously unaffordable "Big Science" facilities. Suddenly, funders had to decide between competing research ideas. At first the most urgent investments – a large telescope, say – were obvious even to outsiders.[3] But this list was quickly exhausted: by the 1920s, foundations were asking prominent scientists to recommend deserving projects.[4] The resulting system fit comfortably with traditions that scientists had used for centuries to decide which papers ought to be published.[5] By the early 1930s, foundations were operating recognizably modern peer-review systems. The federal government adopted similar practices from 1937 onward.[6]

Postwar policymakers elevated this self-governance into a new orthodoxy that allowed "scientists themselves to determine how best to utilize ... resources with as little state interference as possible."[7] Indeed, the NIH made it official policy to "interfer[e] as little as possible with the academic scientists to whom it grants public monies."[8] The result, in the words of one NIH director, was that priorities were determined by a "network" that was "largely self-regulating ... with the same objectives and intrinsic ethic."[9] For their part, academic scientists developed the congenial belief that "the autonomy and right of internal regulation of scientific investigation ... was absolutely necessary for the vitality and success of their enterprise."[10] This ultimately extended self-governance to a wide range of everyday activities, including "controlling the acceptable methods for constructing knowledge, controlling what counts as knowledge through peer review and replication, [and] managing how knowledge is communicated through conference presentations and professional publications." It also meant deciding who would receive resources with "heavy influence on funding through peer review and grant panels, and hiring and promotion of fellow scientists."[11]

At the same time, the Republic of Science had definite limits. Most obviously, Congress continued to set overall budgets, while funding agencies prioritized specific fields and subdisciplines. Still more rules specified how researchers could spend funds, claim intellectual property, practice basic safety, hire employees, and prevent the accidental or deliberate misuse of new discoveries.[12] This last item was particularly important, since it implied that government might sometimes decide to limit scientists' choice of research problems after all. This hardly mattered so long as regulators limited themselves to technologies that had already left the laboratory.[13] But this became untenable after Asilomar conceded the need to regulate individual experiments in the mid-1970s.

The question, as always, was how to square researchers' individual scientific freedom with community-wide regulation. Here, the obvious shadow-of-hierarchy solution was for government to demand self-regulation coupled with the threat of intervention if scientists refused. For this threat to be credible, however, bureaucrats would have had to reject the reigning ideological belief that government intervention would damage research and leave society poorer than before. The result in practice was that grant agencies seldom put much pressure on their recipients. In theory, fears of public backlash might drive them to intervene anyway. However, the defiant attitude of journal editors in our 9/11 and avian flu examples * implies that this effect seldom mattered outside the very largest and most unusual provocations.

---

* The history of UC Berkeley's multimillion-dollar "Synberc" program shows the inertia that even the most activist agencies face when demanding self-regulation. When the author was named the project's "Thrust Leader" for policy issues in 2006, a senior NSF manager remarked that he was happy to see a nonbiologist get the job because the agency wanted

Scientific Merit and Old Boy Networks. Mordern Republic of Science ideology assumes that peer reviewers will allocate resources according to scientific merit. Everyday experience suggests that many scientists accept this principle. At the same time, many do not. While recent data are scarce, a large minority of academic scientists – 40 percent in one older survey – believe that NIH funding decisions have nothing to do with merit. According to them, allocations are instead controlled by old-boy networks, i.e., "a self-perpetuating group of scientists who conspire, perhaps indirectly and at arm's length, to insure that their own work and the work of like-minded scientists receive support, while other proposals go unfunded."[14]

The ubiquity of this belief practically guarantees that "old boy" networks exist and exercise influence at some significant level. But in that case each new funding decision forces academic scientists to choose between competing "old boy" and "merit" networks. They can also change their allegiances over time.

Peer Review as Profit Maximization. As in Chapter 3, we will usually analyze individual scientists' choice between "merit" or "old-boy" networks in terms of profit maximization. Lest this seem arbitrary, we emphasize that the material and psychological resources at stake are substantial. These typically include:

> *Jobs, income, and tenure.* Colleague recommendations play a critical role in most hiring, promotion, and tenure decisions. This crucially includes grants that decide "summer money" worth up to 25 percent of base salary.[15] Peer recommendations similarly decide who receives money through consulting contracts and equity participation in commercial firms. Finally, conference support lets members enjoy lavish accommodations in luxury destinations around the world.
>
> *Job satisfaction and reputation.* Colleague assessments play a critical role in job satisfaction. This includes deciding which experiments will be funded,[16] who receives early access to new ideas, results, and research

---

practical results. In practice, NSF's wishes turned out to be irrelevant because Synberc's biologist leadership set research goals and budgets without consulting me. This pattern became, if anything, even more extreme when anthropologist Paul Rabinow became Thrust Leader. In March 2010, the NSF publicly complained that the collaboration's policy research "appear[ed] to be primarily observational ... rather than proactive and developmental." Rabinow protested that Synberc's scientists had shown themselves indifferent to social concerns, while biologists shot back that Rabinow "had failed to do his job." Six months later Rabinow was replaced by biologist Drew Endy – precisely the outcome that NSF had said it did *not* want. Gollan (2011). Despite this, NSF continued to fund Synberc. Readers can find extended accounts of the affair in Rabinow (2011, 2012).

materials,[17] which papers are published, and which researchers receive awards and recognition.[18]

Healthy communities. Finally, peer review determines the funding of multi-researcher collaborations and entire subdisciplines.[19] Indeed, government grant agencies commonly let large collaborations allocate millions of dollars internally with only limited supervision.

"Splitting the Community." Private politics in our various academic examples almost always revolved around claims that self-regulation would "split the community." Furthermore, members take this danger seriously, often adopting extreme measures (e.g. the cancellation of long-planned votes) to avoid a breach.

The question remains just what members are afraid of. We start from the observation that self-regulation nearly always overrules normal peer review expectations by changing the costs, feasibility, and rewards of doing research.* But in that case, members should reasonably worry that their colleagues will find it easier to ignore peer review a second time. This doubt is destructive. From a profit-maximization standpoint, we expect members to honor principles like scientific merit or old boy cronyism if and only if they can expect reciprocal treatment when the time comes.

The question remains how much a single self-governance vote is likely to damage this trust. In the most optimistic case, members could see self-governance as a one-time departure from normal standards. In that case, both "merit" and "old boy" expectations would proceed much as they had before.† The problem, as our Asilomar example shows, is that it is almost impossible to know whether any single vote for self-governance is motivated by principle or opportunism. We argue below that this suspicion can destabilize networks – and the value they offer to members – for years.

6.1.2. Observed Regularities

Our academic examples share a strong family resemblance. These commonalities include:

Feasibility. The conventional wisdom claims that organizing academic communities is hopeless. However, most of our academic self-governance

---

* For example, atomic physicists were asked to defer publication [Ch. 1.2.1], Asilomar scientists wrote rules that selectively penalized research [Ch. 5.1], and mutations scientists debated a shared database that would have completed with existing projects [Ch. 5.2].
† We neglect the possibility of irrational vendettas and feuds, even though I have heard working scientists mention them as a likely result of "splitting the community."

examples either succeeded (physics, Asilomar) or came close to success (mutations, synthetic biology). Strikingly, this was true even when outside actors like governments and large anchor firms were largely indifferent. We take this as a broad hint that academic communities have considerable power to self-organize.

Politics (A): Participation. Private politics is almost always monopolized by small groups. Indeed, the number of active players in our various academic examples was never more than about 20 percent of the affected community. Some of these established senior scientists who saw themselves as community leaders and/or less vulnerable to controversy. However, participation was also open to younger, less established members who felt strongly about the issue at hand. This is perhaps best exemplified by our Asilomar example, where would-be players routinely forced their way onto key committees by threatening to contact journalists and regulators if the community ignored them.

The fact that so few members are willing to invest the time and energy needed to become active players has deep implications for private politics. Perhaps most obviously, the fact that the average member is usually silent – indeed, may not have formed any deep opinions at all – creates deep uncertainty. It follows that no one can be sure which votes will prevail, how many dissenters exist, or what a vote to "split the community" will cost in practice.

Politics (B) Tactics. Members in our examples almost always tried to avoid clear votes. Instead, they invoked various tactics to limit any potential controversy. These included replacing conventional voting with undefined "consensus" procedures (Asilomar), holding chaotic voice votes while members were leaving the room (Asilomar, mutations), or framing votes in terms of deliberately ambiguous choices (mutations). Dissenters similarly invented excuses to hide their opposition as long as possible (physics, mutations, SB2.0). When ambiguity became impossible, ordinary members frequently abandoned previously announced votes (mutations, SB2.0) rather than risk an open breach with dissenters.*

Interactions With the Broader Society. The chances that government or public opinion will intervene in academic science are generally small. Indeed, even the Asilomar debate received almost no publicity until the

---

* The question arises why dissenters – who also fear voting – should be systematically less afraid of "splitting the community" than the majority. The optimistic answer is that they are not: the fact that the dissenters in our SB2.0 and mutations examples felt so strongly may simply have been luck of the draw. The counterargument is that dissenters occupy the extreme outer edge of community opinion more or less by definition. People who are willing to hold and defend such views against the majority almost always feel strongly.

Box 6.1. Recurring Themes

Community as Social Asset. Peer-review procedures imply that individual scientists often trade places as "allocators" and "recipients" from one transaction to the next. The system is suppose to operate on the principle of scientific merit, but also opens the door to "old-boy" networks in which participants systematically reward each other at the expense of outsiders.

Splitting the Community. Dissenters routinely argue that collective action will "split the community" and that votes are illegitimate. These arguments sometimes deter even large majorities from acting. This strongly implies that "splitting the community" is not just symbolic but also imposes large material costs on members.

Collective Responsibility. The usual Republic of Science rhetoric holds that the decision to conduct or publish experiments is a matter of individual conscience. But in practice, hardly any scientist behaves this way. Instead, even senior figures (e.g., Fermi, Berg) quickly suspend projects in the face of criticism and use the hiatus to solicit support from colleagues. This sometimes leads to further cascades as those contacted reach out to still more colleagues.

Inherently Ambiguous Motives. Practically all self-governance changes the expected costs and benefits of research. Furthermore, those affected seldom have any way of knowing when members who vote for self-governance do so from altruism (e.g., "biosafety") as opposed to self-interest ("not my experiment"). Self-interested votes violate normal Republic of Science expectations and undermine trust going forward.

Participants. Members can conveniently be divided into three categories. *Organizers* run the governance initiative. They decide when votes are held, avoided, or postponed; choose which proposals are developed; and try to expand agreement from small groups to progressively larger constituencies. *Players* participate actively in debates. Their numbers are typically limited since learning is expensive and invites conflict with colleagues. They are disproportionately drawn from senior scientists who are comparatively more able to afford these costs and younger colleagues who are unusually passionate or else directly affected by specific issues. *Followers* refers to the great majority of members who invest a few hours each year in attending meetings. They depend on organizers and leaders for information and often follow their lead. Their passivity – and sometimes delibrate reticence – makes it hard for organizers and leaders to estimate which proposals are popular or would prevail in an open vote.

final meeting.* The fact that community members nevertheless routinely warn against outside intervention suggests that they see even small threats to self-governance as intolerable. This sometimes provides useful political leverage. For example, the Asilomar organizers' unilateral decision to declare a moratorium ensured that the community would face widespread criticism if it failed to adopt voluntary standards. This drastically reduced the chances that the conference would end in deadlock.

6.2. Theory of Networks

Traditional "shadow-of-hierarchy" theory claims that academic self-governance depends on the threat of government intervention. However, this cannot explain our many examples where academic communities act despite official indifference. This suggests that – unlike our baseline commercial model – we should look for the roots of private power within the community itself. Recurring assertions that votes will "split the community" provide a broad hint that this mechanism originates in members' ongoing relationships with each other.

This section develops the intuition by visualizing "the community" as a series of interactions. Section 6.2.1 summarizes what economists have learned about networks where members help each other through nonmarket exchanges that cannot be anticipated or enforced as contracts. Crucially, the value of these networks grows over time as members build trust so that more exchanges become possible. Section 6.3 uses these insights to develop a theory of academic power, emphasizing how votes that "split the community" damage this asset. Crucially, we argue that dissenters will normally suffer much more than the majority. This economic dynamic supplies the de facto constitution for private politics. Section 6.4 describes how academic politics plays out against this background. Section 6.5 uses our analysis to evaluate recent proposals that academic-community politics be opened to nonscientists.

6.2.1. Economics of Networks

Scientific communities famously claim to allocate resources according to scientific merit. But in that case scientists must be allocators one day and

---

* Our synthetic-biology example similarly shows the limits of publicity. Despite the subject's blend of futuristic science and biological warfare, the gap between private and public standards never spilled over from the science press into mainstream media. This may have been a near thing. Reporters for *The Wall Street Journal* and *60 Minutes* contacted the author several times about possible stories. The ETC activist group also considered intervening, but ultimately decided to invest its limited resources elsewhere.

recipients the next. This allows the possibility of repeat trades in which members promise to act on the basis of principles like merit, friendship, or "old-boy" networks.

Conventional markets enforce such promises by contract or simultaneous exchange. But these solutions are not available to academic science where today's allocator cannot be sure what resources he will need in the future, or when, or whether today's recipient will be able to grant them.[20] This means that exchanges cannot be agreed in advance.

The surprise, as economists have emphasized, is that many communities continue to make exchanges anyway. The key is to replace formal contracts with trust, i.e., each party's rational belief that the other will reciprocate when the time comes. Even though the trust literature was originally motivated by pre-industrial bazaar and gift-exchange cultures, economists quickly saw that similar dynamics are ubiquitous in modern economies. Some are illegal, most notably the trust between competitors that supports illegal cartels.[21] Others, as Ostrom famously explained, provide the foundation for private regulation of fisheries, water, pasturage, and other shared resources.[22] Breton and Wintrobe (1983) were among the first to point out that trust networks are ubiquitous in modern universities.[23]

Network Theory. Economists have studied trust networks extensively since the 1980s. While the earliest literature focused on simple, two-person games,[24] it was immediately apparent that the analysis could be extended to complex networks of players. The resulting economics are ably summarized by Jackson (2008). For the most part, this work concentrates on how networks form and/or when and how quickly members' opinions converge.[25] Here, we limit ourselves to the more generic observations that trust must be built up over time and, once constructed, supplies an ongoing flow of benefits to members into the indefinite future. Credible threats to "split the community" put this asset at risk.

"Trust" and "Reputation" Games. Fundamentally, there are two classes of network model. The first is based on trust. At bottom, these theories depend on a kind of "bootstrap" logic: if players expect large rewards from repeated relationships in the future, they will be less likely to betray their partners for a quick, one-time profit.* This is more or less identical to Ostrom's observation

---

* This very fundamental result is formalized by the so-called "Folk Theorem," which holds that "If players are patient (i.e., if the future matters enough), then there exist equilibria of mutual trust." Fudenberg and Maskin (1986). The Folk Theorem derives its name from the fact that it circulated within economics departments for years before ever appearing in print. Its discoverers are unknown.

that accumulated trust is an asset that lets self-governing communities complete larger and more frequent transactions.[26] Perhaps the most important result for our purposes is that trust is constructed over time. Players cannot negotiate large transactions immediately, but must instead build up to them through a series of larger and larger stepping stones.

Crucially, trust can take different forms. For example, members can promise to trade favors (corruption) but also to observe agreed principles (rule of law). In the Republic of Science the former correspond to old boy networks while the latter mean judging proposals according to scientific merit. In practice, both kinds of trust networks can coexist and compete.

The second class of model involves "reputation games." These assume that some, though not all players have inherently honest natures.* Community members must then find out which players are trustworthy and which are not. More specifically, each member estimates her potential partner's honesty and then decides whether to invest in new transactions. Once the transaction is complete, she observes the outcome and updates her estimate. For simple models, this implies that partners who keep promises to each other will build up to larger and larger transactions over time.[27] More sophisticated models recognize that transactions can sometimes fail for reasons beyond the supplier's control.[28] Given that the betrayed partner cannot distinguish accidents from intentional acts, each bad outcome makes future transactions less attractive. But since she understands the possibility of accidents, she could still decide to engage in further transactions (perhaps after a delay) anyhow, so that her partner's estimated reputation slowly recovers.

The point, in both cases, is that trust and reputation are assets which permit transactions that would have been too risky otherwise. Conversely, failure to keep promises impoverishes both sides by increasing perceived risk so that fewer transactions are consummated.[29] These bilateral relations can be readily generalized to networks. At this point, the total stock of trust or reputation between all pairs of players becomes a community asset that determines the total volume of transactions. Collective action votes that "split the community" degrade the asset by damaging relations between dissenters and the majority.

---

* The "reputation" picture recognizes that humans sometimes act from principles beyond simple profit maximization. As Ostrom emphasized, these principles can themselves evolve, most notably through psychological theories in which ethical norms emerge and sometimes even persist in the face of cheating. See, e.g., Ostrom (1990) at p. 88 (evolution of norms), 97–8 (norms and cheating). For present purposes, the distinction seems relatively shallow. The idea that ethical principles are reinforced when community members comply and eroded when they cheat is not very different from the usual logic of trust games.

6.3. Theory of Academic Power

We have argued that the Republic of Science is organized around networks whose members trade places allocating resources to one another. In keeping with Section 6.2, it is natural to expect these exchanges to create trust and reputation networks over time. But in that case, any dispute that weakens these networks impoverishes members. This explains why threats to "split the community" and deference to colleagues are so prominent in the academic initiatives described in Chapters 1 and 5. At the same time, we argue below that "splits" disproportionately punish dissenters. This provides a natural mechanism for collective action.

The trouble, for now, is that social scientists still know very little about how academic science allocates resources. This includes such basic information as how many networks exist, what principles they observe, and how often members switch allegiances. These "details" almost certainly affect how the prospect of "splitting the community" deters dissent and, conversely, empowers collective action. This Section reviews various possible scenarios starting with the simplest case where members belong to a single "Republic of Science" network. We then ask how other, more complex models change the analysis.

6.3.1. The Simplest Model: A Single, Community-Wide Network

We saw in Section 6.1 that academic scientists traditionally claim to allocate resources according to scientific merit. We start with the simplest case where no competing principles exist. The question remains how many scientists would choose to allocate resources according to merit. At one level this decision may seem trivial given that there is only one network to join. However, the material payoff from trust and reciprocity must still offer members more reward than refusing to join any network at all.* To the extent that members do decide to join, they presumably believe that the choice will make them richer. This could be because they believe that rewards allocated according to scientific merit will be more predictable,[†]

---

* For example, members could still decide (a) to save transaction costs by making purely random allocation decisions, (b) act from purely personal and *ad hoc* motives like rewarding cronies, or (c) favor whichever research topics are likely to generate spillovers for their own work. The good news is that modern conflict-of-interest rules mean that members cannot vote to fund themselves. This takes the most destructive possibility off the table.
† We conventionally assume that researchers are risk-averse, *i.e.* that they prefer bargains with predictable payoffs to those which spread the same average reward over a wide range of very large and very small outcomes.

that they themselves possess above-average creativity, and/or that merit-based allocations are more likely to move the field as a whole forward with positive spillovers for everyone.

"Splitting the Community." We now ask what happens in this very simple model when a self-governance vote "splits the community." The starting point, as we saw in Section 6.1, is that meaningful collective action inevitably reshuffles resource allocations compared to the results that "merit" expectations would otherwise have generated. But in that case, members may reasonably suspect that those who voted against the usual merit results will do so again. This undermines their faith in future reciprocity and, with it, any material incentive to remain in the network.

But if dissenters do defect, what happens to the network? So long as the number of dissenters is small, we expect it will survive in something like its original form. That said, it will normally shrink as members exclude colleagues they no longer trust. The net result will be a less capable network, with fewer rewards both for those excluded and those who remain. We argue that this is the real meaning of "splitting the community."

The question remains how these losses will be allocated. Crucially, we argue that the costs of the split fall far more heavily on dissenters than the majority. This generates the punishments that make collective action possible.

To see this, consider the paradigmatic situation depicted in Figure 6.1 where two dissenters break from the wider community. In keeping with our previous discussion, we argue that trust relations that cross the dashed line are damaged by the split and lose value. (Those that do not cross the line have the same value as before.) Readers should note that these impacts are strongly asymmetric so long as the majority greatly outnumbers dissenters. The left-hand side of Figure 6.1 shows what happens to members of the majority. Here Voter "A" has far more links to the majority than to the minority. She therefore suffers relatively little damage from broken trust relations. However, the case is very different for dissenters where the right-hand side of Figure 6.1 shows that almost all of Voter B's links are damaged.

The point, of course, is that the majority can punish the dissenters at relatively little cost to themselves. This supplies the academic analog to the "critical mass" or "voting rule" described in Chapter 3. More specifically, we expect members of the majority to vote for collective action whenever their personal norms and the expected benefits of community-wide rules exceed the personal cost of a split. For dissenters, the decision to defy the majority is painful. Worse, it is potentially unstable since each dissenter who defects to the majority reduces expected costs for the majority and increases punishment for the remaining dissenters. This opens the door to an avalanche

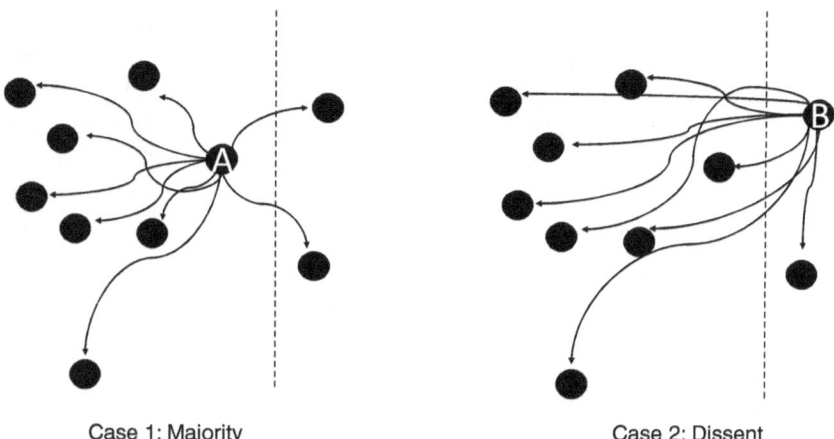

Case 1: Majority   Case 2: Dissent

Figure 6.1 "Splitting the Community."

dynamics in which each defection triggers still more defections. Finally, careful readers will already have noted that our analysis depends on the percentage of broken ties rather than absolute numbers. This suggests that self-governance is feasible even for large communities.

6.3.2. Multiple Competing Networks

We saw in Section 6.1 that the Republic of Science ideal in which a single network dispenses resources according to "scientific merit" is almost certainly too simplistic. Instead, the available evidence suggests that "merit" networks normally compete with "old-boy networks" whose members systematically reward each other at the expense of outsiders. This makes it natural to ask how the existence of multiple, competing networks changes the simple one-network model sketched in Section 6.3.1.

The first and most obvious difference is that researchers must now decide which network to join. For profit-maximizers, this will normally mean choosing whichever network offers the greatest expected reward. In general, we expect researchers who are young, creative, and/or interested in starting new fields to join merit networks. Assuming that good ideas are broadly distributed across researchers, it follows that resources will be widely dispersed across the community.* Merit decisions are also more likely to generate positive externalities for the field as a whole.

---

* The dispersion will be even larger where merit estimates are inherently difficult so that they show large random differences from one reviewer to the next.

But of course, not all researchers fit this profile. In general, we expect researchers who are older, more gregarious, or work in well-established subdisciplines to choose old boy principles instead. The question remains whether we should expect one old boy network or several. For any given network, the profits that *each* member receives will normally depend on how much funding the *average* "old boy" controls. This creates an incentive for very powerful "old boys" to exclude those who are less influential. Economists have shown that inhomogeneous memberships often lead to multiple networks, with the most desirable network consisting of the most powerful old boys, the second most desirable network featuring slightly less powerful boys, and so on.[30]

So far we have assumed that members join whichever network promises the most reward. However, these divisions are unlikely to be static over the long run. For example, a "merit" researcher who gains prestige could become an "old boy" later in life. At the same time, members who switch necessarily forfeit the trust they enjoyed in their former network and must rebuild it from scratch. In normal times, this penalizes switching so that networks are approximately stable.

Finally, old boys, like other community members, have better funding prospects when their field advances quickly. But this is more likely to happen when funds are allocated according to merit. This suggests that merit networks actually subsidize their "old boy" competitors, so that the latter become more profitable and attract more members than they would in isolation.

"Splitting" Revisited. Section 6.3.1 argued that self-governance votes punish dissenters by expelling them from a single network. We expect the same thing to happen when there is more than one network. The difference now is that networks compete and members can switch. This introduces three new effects:

Differential Impact. Self-governance votes will normally affect members' trust in "merit" versus "old boy" networks differently. We therefore expect a net flow away from whichever network happens to be hardest hit.

Reduced switching cost. Members who switch networks forfeit their accumulated store of trust. This normally makes switching costly and discourages defections. But a community-wide split reverses this logic. Given that the split has *already* damaged trust, members contemplating a switch may suddenly find that they have much less to lose. This opens the door to large numbers of near-simultaneous defections.

Fallout. Each member who switches networks reduces the expected reward for colleagues who remain. This invites still more defections. It is natural to assume that these dominoes will continue to fall for several grant cycles – years – before new, realigned trust relations become stable.

6.3.3. Complex Models

Our models so far have been deliberately simple. We close by describing how more realistic – and complicated – scenarios would change our analysis.

Inhomogeneous Links. Economists who study trust networks typically assume that bilateral links between members can have dramatically different strengths, with some links having no strength at all.* In the context of academic science, this means that some dissenters could be more strongly linked to each other than they are to the majority. (This might happen, for example, if they are coauthors or belong to the same laboratories.) The fact that conflicts in our Asilomar example often broke out along disciplinary lines suggests that these asymmetries are important.

To the extent that asymmetries do exist, the community's power to punish dissenters will often be attenuated. For example, dissenters who are strongly tied to each other in tight, self-sufficient islands will often be able to do science with or without the majority's consent. This hints at a complicated politics where the decision to dissent occurs collectively within each island, with locally powerful figures like lab heads exercising disproportionate influence.

Sequencing of Disputes. Our mutations example presented the very simple case in which two dissenters faced off against an entire community. However, more complicated institutions and timing are possible. This is nicely illustrated by our SB2.0 example, in which the dissent took place entirely behind closed doors so that it never reached an open meeting. Whether this changed the final result depends on one's theoretical prejudices. On the one hand, it is possible to imagine the closed-door group concluding (a) that dissenters were prepared to defy the full community, and (b) that the community would then back down. If so, the private decision to derail debate did nothing to change the outcome while neatly avoiding a public embarrassment. On the other hand, one can equally imagine the closed-door group wanting

---

* Trust (reputation) links are also asymmetric, i.e., A can repose more trust in B than B does in A. Breton and Wintrobe (1982) at p. 65.

to decide the question among themselves rather than share power with outsiders. In that case, the closed-door group would itself become a kind of old-boy network in which members trust each other to vote in ways that avoid offending any single member unduly. This could systematically cripple collective action to the extent that dissenters possess stronger ties within the closed group than they would in an open meeting.

Uncertainty. We have argued that players seldom know how deep – or costly – a split will be. Assuming normal risk aversion, we expect the majority to fear the worst and be deterred more often than it would be under full transparency. But in that case, organizers should welcome any tactic that reduces uncertainty short of a vote. This recalls the crucial Asilomar moment when Prof. Botstein called for a nonbinding straw poll that ultimately revealed that the number of dissenters was much smaller than most members expected.*

Anonymity Rules. Figure 6.1 ignores the common – if imperfect – rules that shield the anonymity of referees.† This could lead to more complex models in which members place their trust in networks without knowing how individuals vote in cases that affect them. For now, it is unclear how important anonymity really is. For example, most job candidates know perfectly well who supplies their recommendations. The rise of Big Science consortia in which dozens and even hundreds of scientists allocate funds internally[31] suggests that anonymity is less important than it used to be.

Commercial Interactions. Finally, we have so far assumed that member rewards are determined entirely by trust relations. However, the Republic of Science is embedded in a wide array of government ("shadow of hierarchy") and commercial ("shadow electorate") institutions. This suggests that grant managers and biotech executives may sometimes bribe or threaten members to change their votes. Similarly, our mutations example shows how community-infrastructure projects can generate new resources that did not

---

* One might, of course, object that a straw vote could harm trust just as easily as a real one. This might not be true, however, in the typical situation where everyone understands that members' positions are tentative and can change dynamically depending on how many dissenters raise their hands.
† Most academic readers will know that referees' identities can often be inferred from the internal evidence of their comments, especially including the authors and arguments they rely on. Furthermore, referees in other settings (e.g. grants panels) often know each other's identities even if the applicant does not. This gives each referee ample opportunity to observe her colleagues and, if necessary, recalibrate her estimates of how often that colleague honors "merit" versus "old boy" principles.

exist before. This means that they will sometimes be able to pay opponents to work on projects that they would otherwise oppose.

6.3.4. Enforcement

Commercial self-regulation is almost always enforced by anchor firms outside the community itself. The mechanisms for enforcing academic standards that originate inside the community are necessarily more complex.

Community Enforcement. The most obvious way to enforce rules is to break off trust relations with those who violate them. This, however, is costly to both sides so that enforcers may be tempted to look the other way. One partial solution is for the community to publicize its rules as a commitment strategy. This makes infractions more visible to the general public if and when members fail to enforce them. Alternatively, members can establish institutional watchdogs that specialize in monitoring and announcing violations. The problem, as we saw in Chapter 1, is that this kind of third-party audit structure is easily corrupted.

Where the community does decide to act, it is seldom necessary to sanction more than one violator at a time. This implies that members can afford to be harsh. The physics community's postwar ostracism of Edward Teller for testifying against J. Robert Oppenheimer shows how strong such sanctions can be:

Teller was an outcast ... There was a famous incident at Los Alamos soon afterward, where people refused to shake his hand. And he was shaken, and went back to his room, and was extremely upset. And lost his friends. Was a pariah in the physics community, really, ever since then.[32]

Government Enforcement. A politically stranger enforcement scheme, following our Asilomar example, is for officials to encourage the community to self-govern and then codify the results in conventional regulations. Alternatively, government could deny grants to experiments that violated the private standard. The obvious question, in both cases, is why officials would not simply write and impose their own rules at the outset. One answer, following our commercial examples, is that scientists will usually find more cost-effective solutions than any outsider. Additionally, delegation provides a measure of political discovery since communities have a built-in incentive to develop solutions that minimize members' unhappiness.

The downside, of course, is that delegating power to a self-interested community invites minimalist solutions. We saw in Chapter 2 that anchor firms like Unilever and Home Depot avoid this dynamic by encouraging suppliers to compete to see who can deliver strong standards. Government could create similar incentives by announcing a preference for, say, grant applications that promise tougher security standards. The question remains whether grant authorities should let members develop these standards collectively. On the one hand, common standards let members share development costs and lead to more homogeneous practices. On the other, collaboration could lead to a de facto cartel where every member commits to offering exactly the same precautions.[33]

The other great weakness of government enforcement is that it is limited by the reach of U.S. law. This suggests that additional methods may be necessary overseas.* That said, Western funding agencies often follow the NIH's lead, especially for standards that originate in international meetings on the Asilomar pattern.

Journal Editor Sanctions. Finally communities can trust enforcement to journal editors. One obvious objection is that an agreement by lead journals – say, *Nature, Science*, and *Cell* – might not prevent dissenters from publishing in less prestigious venues. Still, this is already in progress since the prospect of publishing in an obscure and low-prestige journal will often reduce the temptation to violate standards in the first place. Furthermore, mid-ranked journals will likely receive much more criticism if higher-ranked journals have already publicly refused to publish. Finally, lower-ranked journals would face special pressure to conform, given tight university library budgets and widespread criticism that many journals are not worth the subscription price.[34]

6.3.5. The Limits of Academic Power

We have presented a series of simple models in which dissenters decide that it is better to accept self-governance than to break off relations with the majority. But trust and reputation networks are only part of the story. This section asks whether and to what extent institutions like tenure and outside commercial income insulate members from trust and reputation sanctions.

Tenure. We argued in Chapter 3 that commercial firms that defy community standards can find themselves earning negative profit. This is in itself a kind of death penalty. The difference in academic life is that most faculty are both salaried and protected by tenure. If they want to, dissenters can

---

* This should not be overstated. Many foreign researchers subject themselves to U.S. rules by joining NIH-supported projects.

persist indefinitely. The empirical question remains how often the prospect of diminished resources nevertheless persuades scientists to shift their research to other, less controversial topics.

In practice, relatively little work has been done in this area. However, modern accounts generally agree that the effect is substantial. These include anecdotal evidence that increasing regulation has persuaded many scientists and institutions to abandon research into biological weapons organisms "to some degree already"[35]; that even established laboratories sometimes close down when researchers fail to attract new grants[36]; and that some young researchers stop doing research entirely.[37] Unfortunately, the only quantitative study I know of is ancient: survey evidence from the early 1970s[38] shows that roughly one-sixth of academic medical researches who were denied grants had to change jobs, and that this rate doubled for faculty without tenure.[39]

Reduced Resources. A second, harder limit is set by the fact that many experiments require significant resources. Contemporary scholars usually argue that cutting of resources means that many experiments will never be conducted at all.[40] Simple experiments can often be done using borrowed equipment and materials "bootlegged" from other grants However, the feasibility of these workarounds declines steeply for complex projects. This notably includes many of the most worrisome experiments, including projects that simultaneously insert large numbers of genes into existing genomes or aim to create engineered organisms that can survive outside the laboratory.[41] During the 1970s, at least, the deterrent effect seemed to be large. Half (49 percent) of the investigators who employed technicians had to lay off at least one when their grant application was denied – a significant loss of capability.[42] For scientists who stayed in the field, only about one-tenth (13 percent) persisted in doing at least part of the proposed work from their own pockets, using borrowed equipment, and resources diverted from other grants.[43] The number who stayed with the original idea fell even further to just 7 percent after several years without support.[44]

Outside Commercial Income. Asilomar veterans sometimes complain that today's scientists have many more ties to biotech companies,[45] and worry that this could "complicate" future "self-scrutiny."[46] However, we argued in Chapter 3 that many companies face shadow electors and the threat of backlash. Given that corporations typically have so much at stake, they can afford to pay individual dissenters to join the majority after all. Conversely, payments to the majority could outweigh whatever deterrence dissenters can generate by threatening to "split the community."

6.4. Theory of Academic Politics

We have argued that trust relations in science define a voting rule that lets large majorities overrule dissent at little cost to themselves. As in our commercial discussion, we expect politics to unfold against this rough-and-ready "constitution."

6.4.1. Internal Politics

We have already explored private commercial politics in Chapter 4. Here, we focus on the special features that make academic politics distinctive.

Limited Information Budgets. We start from the observation that individual members' budgets for obtaining information and negotiating agreements are painfully limited. It is easy for companies to study issues when million-dollar payoffs are at stake. By comparison, academic scientists' interest in particular issues seldom extends beyond their own labs.

Political debate is further limited by the difficulties in communication. The fact that the average academic community consists of several hundred members – 150 to 650 in our examples – sets a particularly awkward scale for private politics. On the one hand, it is much too large for face-to-face negotiation on the StarKist pattern.* On the other, it is too small to support the elaborate investments in rules and institutions that smoothed information sharing and negotiation for our commercial fisheries, coffee, and lumber examples.

A corollary of members' limited information budgets is the near-impossibility of predicting which coalitions and standards are feasible. Before the actual meeting, most members will have only tentative and changeable opinions. Even if a poll were taken, there is very little to measure. Worse, members know that silence is strategic. We have argued that academics centrally depend on trust relations, and personally expect to suffer each time those relations are damaged. This implies that the average member will almost always prefer to let someone else argue his position, standing up only when no one else does.[†] But this leaves no way to count heads: if the current crop of dissenters was removed, others might immediately spring up to replace them. Or they might not. There is no good way to tell.

---

* The fact that active players seldom number more than 10–20 percent of any given community may sometimes reduce this to a manageable number.

† The case is very different in commercial settings, where discussions tend to be frank. The reason, almost certainly, is that players already compete so fiercely that they have little to fear by offending.

Organizer Strategies. As in our commercial examples, organizers possess large amounts of time and energy to collect and disseminate information. They also have the usual speaker's prerogative of controlling the timing and content of votes. Following our mutations and Asilomar examples, we expect them to use these assets to develop solutions that initially command consensus in small groups, thereafter expanding the circle until the full membership is ready to adopt the standard.

Given members' limited information budgets, organizers will often find it hard to convey even the most basic and undisputed facts. Much of this communication will require labor-intensive bilateral conversations. Sometimes, however, organizers may be able to save effort by hosting group conversations in the form of small planning meetings, study groups (e.g., Asilomar, mutations) or webcast town hall meetings (e.g., synthetic biology) before the full membership meets. That said, these mechanisms are only likely to work where the wider membership is willing to accept the smaller group as its trusted intermediary.

Finally, communicating *political* facts presents special problems. As we saw in our mutations and synthetic-biology examples, it is not enough for organizers to convince members that a particular standard makes sense in the abstract. They must also know what price they will pay if and when the community "splits." Botstein's suggestion at Asilomar that members hold a nonbinding straw poll provides one of the few obvious tactics for previewing votes before they are actually taken.

Player Strategies. One consequence of limited information budgets is that the average member must be persuaded to vote for collective action based almost entirely on what she learns at the meeting itself. This favors tactics in which opponents try to overwhelm members by forcing them to sort through deliberately misleading facts and arguments. Given that members' information budgets are barely adequate to begin with, this can easily lead to paralyzing uncertainty.* A second strategy is for dissenters to underscore both the probability and cost of a break by showing anger. As Elinor Ostrom has noted, "Confrontational strategies raise transformation costs sharply."[47]

Probably the biggest oddity of academic politics compared to the commercial kind is that the split itself can be ambiguous. For example, most of the Asilomar dissenters eventually softened their "no" votes with more

---

* This tactic was particularly prominent in our mutations example, where opponents claimed that better deals were possible up to and including new faculty chairs for members. None of these have ever materialized.

ambiguous after-the-fact comments.* This ambiguity was almost certainly deliberate: had Watson genuinely wanted a split, he would have insisted that the junior workers at his laboratory follow him into the minority. In these very special cases, the fact that a few elites chose to dissent is most naturally understood as a symbolic reminder that the community should not take their support for granted.

Prospects. The question remains what if anything can be done to improve internal politics so that it converges more reliably on Downsian outcomes. Here external competition seems unlikely, since there is only one consumer (government) to arbitrate the contest. That said, one can at least imagine the NIH awarding "plus factors" to grant applications that offer better biosecurity, say, than competing proposals. Another possibility would be to increase organizers' information advantage by supplying them with paid staff. This intervention seems particularly natural where sponsors like the March of Dimes encourage communities to organize collective action.

Finally, it is hard to know how much the existing weaknesses of academic self-governance reflect members' lack of experience. Provided that self-governance grows to the point where members start to hold regular, repeated votes, supporting norms and institutions are bound to emerge automatically. This makes it natural to hope that members could eventually develop new trust relations to suppress Asilomar-style fears that the majority will try to "solve" its public relations problem by imposing complicated regulations on the minority. As Ostrom reminds us, "Fears that a minimal coalition of winners will impose costs on the losers makes them more likely to oppose. Therefore, norms against such behaviors make rule changes less costly."[48]

6.4.2. Mass Politics

We have argued that commercial self-governance often plays out against strong shadow electorates. The choices for the Republic of Science are more limited. According to the traditional shadow of hierarchy conception, government officials are expected to intervene if self-regulation proves inadequate. In fact, they have historically shown little or no appetite for taking this step, and indeed the whole rhetoric of the subject warns that such interventions would do more harm than good.

Theories that depend on the threat of backlash – whether driven by outraged officials, Congress, or the public at large – are more historically

---

* The fact that the dissenters were current or future Nobel Prize winners made it even less likely that they would suffer permanent ostracism.

plausible: while they seldom occur, many scientists in our examples seemed genuinely worried about the possibility. The problem, as with all backlash theories, is that the threatened overreaction is only effective when it is occasionally carried out. This promises to make society worse off on net.

Strangely, the most hopeful scenario involves academic scientists' frequently criticized links to commercial firms. Since start-ups and large drug companies seldom suffer from quality competition, the possibility of shadow electorates is remote. However, their executives often have strong personal beliefs and/or sound business reasons to fear backlash. Best of all, biotech companies often operate extensive consulting networks to detect and exploit the latest developments in academic science. This suggests that, contrary to the usual Republic of Science arguments, they may possess enough detailed knowledge to intervene without damaging the basic enterprise.

6.5. Proposed Reforms

Today's peer review methods for allocating research funds date from the 1940s, when the Republic of Science had helped win the Second World War and seemed poised to do similar things for consumer goods and human health.[49] The public would never again be so trusting. Increasing skepticism since the 1970s has generated proposals to inject nonscientists into the Republic of Science's debates.[50] In principle, there are four ways to do this.

Experts. Following Asilomar, Watson enlisted academic experts from the humanities and social sciences to explore the ethical, legal, and social implications of genetic engineering. His hope was that this would lead to a revised social contract.[51] The European Commission later adopted a similar approach.[52]

Relying on experts is supposed to "de-politicize" science so that consensus solutions can emerge.[53] Probably the main advantage is that participants already speak a common language. This presumably facilitates communication and keeps members' costs of acquiring and processing information to a minimum. This implies a more predictable politics with less chance of deadlock. However, there were also downsides. First, trained ethicists' insights are not obviously superior to ordinary lay opinions.[54] Second, there is no reason to think that academic opinion reflects the broader society. Indeed, many academics had been socialized into the same circles as biologists themselves. Finally, paying academics to participate necessarily reduces their willingness to walk away from collaborations they disagree with and could lead to capture.[55] This dilutes any benefits the mechanism might otherwise have in signaling responsible behavior to the broader society.*

Individual Citizens. The polar opposite of expert meetings is to invite communication between scientists and average citizens. Many countries, especially in Europe, have established biomedical ethics councils that include scientists, citizens, and politicians.[56] These are organized in various ways, including citizen juries, panels, consensus conferences, planning cells, deliberative polling, focus groups, surveys, public hearings, open houses, advisory committees, community planning, and referenda.[57] Some try to set policy while others are content to identify uncertainties and options.[58] The stated goals range from surveying public opinion, consultation, direct democracy, and procedural justice to simply avoiding "protest and reduced trust in governing bodies."[59]

At one level, just agreeing to talk provides a powerful signal that scientists have nothing to hide. However, this is limited by the fact that members of the public almost always lack the time to learn and process information in depth. This is especially true since the conveners can usually use their control of the agenda to frame public participation in ways that favor preconceived outcomes.[60] In theory, these drawbacks could be overcome by conscripting and educating randomly selected members on the pattern of jury duty or, more realistically, paying citizens for their time. However, this process would itself be vulnerable to the charge that education could inadvertently indoctrinate citizen representatives with scientists' normative views alongside objective information. To the extent that members are paid, participation could also create conflicts of interest. This would necessarily erode signaling and public trust. Unpaid participation, on the other hand, would skew participation to members who were unusually passionate, raising the usual question about how faithfully such people represent average (and usually latent) attitudes within the wider society.

NGOs. So far we have emphasized the tension between enlisting conflicted but well-informed experts and sincere but unsophisticated members of the public. This suggests a middle ground based on enlisting NGOs and other trusted intermediaries. The fact that firms routinely delegate power

---

* This reliance on experts reached a kind of reductio ad absurdum when synthetic biologists staged *invitation-only* meetings with other academics "to reassure the public." Garfinkel et al. (2007) (reporting results of three invitation-only meetings to identify "options" for synthetic-biology policy).

to such bodies in our commercial examples strongly suggests that academic communities could too. After all, there is far less hostility and suspicion to overcome. While there is little hope that NGOs truly represent the wider population,[61] the public might not care if fanatically suspicious groups nevertheless declare themselves satisfied.

To some extent, our categories are a false choice: policymakers can and probably should develop all three. Here the most natural strategy would be to develop consensus first among academics, expanding it successively to activists and random members of the public. The question remains how much power they receive. This is presumably measured – as in our other examples – by what would happen in the event of walkout. The obvious difference, however, is that there is no shadow electorate to respond. Instead, a walkout would have to elicit action from (a) funding agencies that treat outsiders as a warning signal that it is time to intervene over scientists' objections or (b) public or congressional outrage leading to backlash in Congress or the funding agencies. On the historical evidence, neither of these channels appears to be particularly responsive.

Corporate Influences. Finally, we have argued that many industries are careful to respect the wishes of consumer and/or employee shadow electorates. It is natural to think that these same preferences could ultimately influence academics through their participation in start-ups, consulting agreements, and contract research. For now the evidence is limited. That said, we saw in our artificial-DNA examples that firms played a leading role in putting security issues on academic biology's agenda.* This suggests, but does not prove, that corporate involvement could indeed provide a useful vehicle for injecting widely-held values from the broader society.

---

* There was also persistent speculation that some companies pressed their academic partners to shelve their private opposition to regulation.

PART IV

LEGITIMACY, LAW, AND POLICY

7

Legitimacy

We argued in Parts II and III that self-governance is feasible and effective. By comparison, our understanding of whether it is normatively acceptable ("legitimate") remains a work in progress. Private transnational governance, even more than its domestic predecessors, remains "a conceptual hodgepodge"[1] without perceptible consensus.[2] The central difficulty is that private politics has no mass elections. As Meidinger remarks, "[I]f these competitive supragovernmental regulatory systems further democracy, they do so in ways that do not fit very comfortably with received models of democracy."[3]

Apart from their inherent interest, legitimacy questions have important practical applications. We saw in Chapters 2 and 5 that self-governance bodies often devote extensive debate to whether their procedures and even existence are legitimate. This drains members' already-limited budgets for finding substantive solutions. At the same time, different groups have a way of rediscovering the same legitimacy arguments over and over again. This suggests that it would be better for society to convene some experts' group to work through the issues once and for all in whatever depth is needed. This is a natural role for scholars.

This section reviews the existing legitimacy literature to collect what scholars have learned about the main issues and arguments so far. Unlike most earlier scholars we do not try to address self-governance in its full generality. Instead, following earlier chapters, we focus on the specific institutional models that have emerged since the 1980s. We will see that this narrowed problem is significantly more tractable.

We proceed as follows. Section 7.1 presents a brief introduction, emphasizing the need to find deeper principles beyond ad hoc recommendations. Section 7.2 asks whether private government can be legitimate even in principle. Sections 7.3 and 7.4 focus on modern democratic theory, emphasizing how shadow electorates and transparent debate could plausibly compensate for private bodies' inability to conduct elections. Section 7.5 addresses intellectual shortcuts that treat legitimacy as something that can be measured empirically or borrowed from traditional governments.

7.1. The Debate So Far

Private standards bodies have existed for more than a century. Despite this, there was almost no "attempt to conceptualize theories of representation" as recently as the 1970s.[4] The reason, as Russell[5] explains, is that almost all of today's industry self-governance traditions descend from standards-organizer Paul Gough Agnew's pioneering work in the 1920s. But Agnew did not think very hard about the roots of legitimacy. Instead, he tried to acquire it the easy way, by persuading governments to join and endorse private standards bodies. This idea was later discredited by the experience of government/employer/worker councils under Fascism.[6]

Meanwhile, private standards bodies tried to meet public criticism by constantly adding rules. The result, by the 1970s, was an "essentially … shotgun approach to a concededly difficult representation problem," producing "a hodgepodge" of unreflective practices "[l]ike hex signs on Amish barns."[7] This incoherence is if anything even more pronounced in modern documents like ISEAL's widely imitated "Credibility Principles" and "Code of Good Practices."[8] These list multiple process goals ("sustainability," "engagement," "transparency" …) and rules, most notably that decisions should be made by less-than-unanimous "consensus."[9] But ISEAL never even tries to answer basic questions like how much consensus or transparency is enough. This deliberate vagueness has debased the discussion to the point where even scholars sometimes find themselves using terms like "legitimacy" and "illegitimacy" as "general labels of approval and disapproval" without "any precise meaning."[10]

Recent scholarship seeks to replace these folkloric approaches with clear, first-principles theories of when private self-governance is legitimate, even while conceding that "convincing democratic theories for the global sphere … [remain] lacking."[11] The good news is that scholars have compiled an exhaustive list of possible principles. The rest of this chapter compares these threads against the specific self-governance models described in Parts II and III.

7.2. Is Legitimacy Possible?

We begin by asking whether legitimacy is possible even in principle. This section identifies two of the deepest objections: that New Self-Governance communities hardly ever include all affected parties, and that they often generate skewed outcomes that systematically favor some members over others.

7.2.1. Defining the Demos

Liberal democratic theory typically starts with the concept of communal identity or "demos."[12] Conventional governments almost always define this geographically. However the New Self-Governance lacks this option because "Only a very small minority of people" currently "identify and communicate with other people on a global basis."[13] Instead, rich-nation activists and consumers end up imposing their standards on developing-world populations that often have strikingly different values. As Meidinger points out, this makes them "subjects and not citizens of the regulatory programs." This neo-imperialist objection "is probably the most probing test of the democratic claims of competitive supragovernmental regulatory programs."[14]

The traditional answer to such arguments is that the demos can emerge over time when members who could theoretically leave decide to stay.[15] But since the New Self-Governance is coercive, leaving is not really a choice.[16] Democratic theory traditionally evades the objection by arguing that small minorities should not be allowed to impose their will on the majority. However, such arguments are much less convincing for the New Self-Governance, where those coerced almost always outnumber the rich-nation consumers whose purchasing power enforces the rule. At this point, appeals to conscious choice become more or less untenable.

This leads some scholars to argue that New Self-Governance organizations are legitimate if they impose values that citizens in developing nations *would* want if regulation was effective.[17] The trouble is that simple survey polling cannot settle the matter since most people in the developing world have never experienced Western-style regulation, much less formed policy preferences. But in that case, we should at least insist that their opinion be measurable in the future. This seems to authorize a kind of democratic imperialism in which private governments are allowed to operate long enough to gain local populations' approval, without any clear standard for deciding how long these "temporary" interventions should last.

A better answer modifies the usual "Veil of Ignorance" argument* to define the demos itself. Start by imagining that every human on the planet was asked to approve a constitution *without knowing her future place in society*. Then we can immediately ask which kinds of citizens should have a voice in solving particular problems. The usual Veil of Ignorance argument holds that profit-maximizing members agree to constitutions because collective action opens life opportunities that they cannot achieve otherwise. Applying this logic to communities implies that citizens would want a demos large enough to include everyone potentially affected by the problem. New Self-Governance organizations routinely achieve the required scale, with the demos that addresses fisheries, for example, by including the populations that consume *and* harvest fish.

7.2.2. Truncated Outcomes

Governments based on physical force can implement any constitution they want. Private power, by comparison, requires a critical mass of anchor firms.† This suggests that private power cannot yield outcomes that consistently frustrate the anchor firms' preferences.

The question is whether citizens operating under a Veil of Ignorance would ever choose such a thing. Skewed or not, the possibility of private governance increases the range of possible collective action. The fact that private governance systematically favors anchor firms might not matter if other actors benefit as well. In this sense, private power expands the number of potentially democratic outcomes compared to what existed before.

Finally, Veil of Ignorance arguments suggest that legitimacy often requires constitutional guarantees that limit the majority's power to impose outcomes on minorities. Many scholars go down this path by arguing that New Self-Governance bodies are more legitimate when they give citizens of the developing world a say in outcomes that affect them.[18] The question is how to make these participation guarantees binding. Here the key

---

* Veil of Ignorance arguments ask what rules voters would agree to if they did not know what social status they would occupy and hence how the rules would affect them as individuals. Buchanan and Tullock (1974 [1962]) at pp.77–8; Rawls (1999 [1971]) at p. 118. The basic insight is that utility-maximizing individuals will always agree to collective action when it opens possibilities that would not otherwise exist for them. This is true even when collective action is conditioned on restrictions that limit how those possibilities are used.

† We have argued that this initial condition can be modified to the extent that anchor firms later find it in their interest to delegate power to others. But this is only a partial answer since delegation is not binding. If they want to, anchor firms can always reverse themselves and take back power.

is convincing the majority that it is more important to honor constitutional safeguards than to win every short-term policy quarrel. This instinct is common in traditional governments, most spectacularly in the United States, where many citizens accord the Constitution more respect than the Founders themselves did.[19] Private political systems, being newer, have had much less time to develop such traditions. In practice, the public often seems to prefer drumhead justice,* although it is worth noting that similar things have happened in traditional governments, including, notoriously, the internment of Japanese-Americans during World War II.[20]

7.3. Representative Democracy

Even assuming that legitimacy is possible, we must still find institutions to implement it. For the most part, the self-governance literature starts from democratic theory[21] with a sub-focus on "representative"[22] or "aggregative" models in which "the people" elect representatives who vote on laws.[23] This faces the obvious objection that private bodies do not hold mass elections[24] and have little chance of doing so.†

The question then becomes whether elections are the one and only way to create legitimacy. This is a serious question, given the hold that elections exercise over modern imaginations.‡ In practice, very few authors insist on

---

* Anecdotally, at least, private bodies seem quick to jettison internal protections when the public is incensed. In 2012, for example, the U.S. National Basketball Association (NBA) received nearly universal praise for forcing out an apparently racist owner. The fact that at least ten fellow owners announced that they would support expulsion before the accused had had a chance to defend himself under the NBA's own rules was universally ignored. Simpson (2014).

† Web-based votes could conceivably change this. For a short summary of the (considerable) technical challenges, see VerifiedVoting.org (n.d.). For now, the closest examples involve sports. Winning candidates for Major League Baseball's All-Star Game regularly garner 1.5 million votes, with the caveat that individual fans are allowed to cast up to 35 votes each. Newman (2016). The ability to vote more than once might even be defensible. Provided that the cost of voting is high enough, repeat voting could substitute for intensity of feeling – something that conventional majority rule notoriously lacks apart from second-order mechanisms like logrolling. Downs (1957).

‡ Che Guevara famously remarked that revolution is doomed to failure "Where a government has come into power through some form of popular vote, fraudulent or not, and maintains at least an appearance of constitutional legality." Guevara (1961). Even so, the modern preference for elections is remarkably recent. Before the eighteenth century, most theorists saw democracy as so preeminently corruptible that they preferred to find "the will of the people" elsewhere. Echoes of these views lasted well into the twentieth century. Waldman (2016); Hobson, (2009) at p. 632 ("The now widespread agreement over the normative desirability and political legitimacy of democracy is noticeably different from the historically dominant understanding that regarded it as a dangerous and unstable form of rule which inevitably led to anarchy or despotism").

elections and then usually as a kind of devil's argument or semiempirical observation.[25] This makes it natural to ask whether our "shadow electorate" concept can fill the gap.

7.3.1. Shadow Electorates

We have argued that quality competition forces corporate executives to pay attention to consumers' policy preferences for much the same reason that elected officials pander to polls. For reasons already noted in Chapter 4, shadow electorate membership closely mirrors the de jure one, particularly when we consider that the composition of the latter often varies significantly from one election to the next. The bundling of issues is also similar. The question remains how close these resemblances should be before private bodies are considered "legitimate." This is unlikely to have any very principled answer. However, this situation is no worse than conventional legitimacy theories, which must similarly decide when the messiness of real democracies differs "too much" from the theoretical ideal.

7.3.2. Legitimacy of Private Actors: Executives, Legislators, Delegees, Activists

The shadow electorate does not govern directly. Instead, we have argued that its preferences are elaborately mediated through complex networks of executives, private legislatures, activists, and, sometimes, external politics. The net impact of these institutions is not obvious. On the one hand, formal rights mean very little if the shadow electorate does not understand its choices. Institutional structure and competition can potentially fill this gap. On the other hand, all structures are imperfect and allow individual players to frustrate the outcomes that a fully informed shadow electorate would prefer.

Executives. On our limited state of knowledge, it is at least reasonable to think that New Self-Governance executives possess comparable discretion to agency officials in iron triangles or elected politicians facing reelection.* Indeed, the fact that so many activists choose to pressure corporations provides persuasive evidence "that business is far more susceptible to pressure

---

* The single greatest weakness of such arguments is that retiring ("lame duck") politicians who do not face reelection have nearly unlimited discretion to do what they like. The situation is, if anything, worse for modern politicians who frequently try to build "legacies" as a monument to their own careers. The saving grace, especially in the American system, is that separation of powers places limits on what even the most powerful individuals can accomplish.

than government."²⁶ Since executives are unelected, shadow electorates cannot select them. Trust can, however, be inherited where executives are recruited from established third-party organizations, or else earned over time as executives compile track records as philosopher-kings.

Legislators. We start by acknowledging that private legislators suffer deep disabilities compared to their public counterparts. This is partly due to the fact that our "shadow electorate" votes on policies rather than individuals,²⁷ preventing would-be representatives from campaigning on promises to implement what voters *would* want, given complete information.²⁸ Worse, traditional "one representative, one vote" prescriptions have no clear meaning in bodies whose memberships are self-selected from enthusiasts. Whether these groups "represent the 'public interest' or only their own private interests is highly contested."* It follows that any scheme that lets all representatives vote in a single chamber, with or without weighting, cannot help making "invidious ideological choices among groups."²⁹

Private standards bodies implicitly acknowledge these difficulties by adopting collectivist models that replace simple majority rule with multiple "chambers," each of which houses a particular interest group that must separately agree to action.³⁰ But this leads to the further difficulty that the chambers themselves are seldom unanimous.† The vaguely defined concept of "consensus" is meant to fix this. However, it is hard to fit "consensus" within standard democracy theory. We argued in Chapter 4 that private legislatures can provide *stochastic representation* in the sense that a large enough supermajority of representatives reliably demonstrates a simple majority in the wider society. While the size of this supermajority is unknowable ex ante, we expect the correct number to emerge from experience: if external challenges invariably fail, the working definition of consensus is probably adequate. A variant argument is that "consensus" is defined by the number of walkouts that would unacceptably cripple the private body. We

---

* The situation is even worse where members include entities, so that some members speak for thousands while others represent only themselves. See, e.g., Abbott and Snidal (2009b) at p. 61 ("NGOs represent their own private values. This is only part of society's interest and in the developing world raises issues of accountability and paternalism"); Grant and Keohane (2005) at p. 38 ("In practice, few NGOs have well-defined procedures for accountability to anyone other than financial contributors and members – quite a small set of people"); Dixon (1978) at p. 37 ("There is the additional complexity that the so-called public interest groups are normally not active membership groups at all but small offices with impressive names resting on government grants, foundation grants, awards of counsel fees in successful class action suits, and the like").

† Even if they were, unanimity itself is poorly defined since dissenters would almost always be included in a sufficiently expanded chamber.

have argued that these groups' ability to negotiate power *within* the private bodies ultimately depends on the outcomes that they could hypothetically achieve by going *outside* those bodies to petition more conventional bodies' institutions like courts, government agencies, and the press. To the extent that these outside institutions have correctly determined how much influence any particular group "ought" to have, it follows that the group's influence within the private body should automatically possesses a measure of derived legitimacy.

Delegees. Even where anchor firms are answerable to shadow electors, they often do not know enough to impose standards on suppliers. We saw in Chapter 2 how this led to a partial delegation of power in our food example and a near-total delegation in the case of artificial DNA. By analogy with traditional "shadow of hierarchy" arguments, the most natural interpretation is that anchor firms have rationally traded detailed control for the prospect of more effective and efficient standards. To the extent that anchor firms' power is itself legitimate – for example because executives feel constrained by shadow electorates or fear backlash – delegations to exploit that power more effectively should also be acceptable.

However, there is also a deeper argument for delegation. Veil of Ignorance arguments suggest that when humans band together to achieve power over nature, they also accept limits on how individuals can use that power. It follows that even small groups have a right to self-regulate unless and until some larger demos overrules them.

7.3.3. Institutional Designs

So far, we have argued that private self-governance is legitimate in principle. Whether it is also legitimate in practice depends on institutions. While many scholars treat transparency, consensus, and broad representation as self-evidently beneficial, it is easy to imagine counterexamples where they might be harmful.* We should therefore look for situations where transparency really does improve matters. Our analysis in Chapter 4 suggests two such situations. First, shadow electorates are systematically disenfranchised where transparency is poor.[31] In principle, at least, better institutions can fix that.[32] Second, transparency can also improve the internal politics of governance bodies. Here we expect institutions that reduce information costs to make politics more predictable while bringing average outcomes closer to Downs's predictions.

---

* For example, "transparency" makes it simpler to know when illicit agreements are being honored and, to that extent, notoriously helps to stabilize illegal cartels.

That said, transparency and due process are only means to an end. We have argued that external competition can sometimes generate information more efficiently than the most virtuous conventional institutions.[33]

7.3.4. Comparative Democracy

Scholars usually frame legitimacy as a binary choice: systems are either legitimate or they are not. But this is only a convenience. The more careful answer is that we should want the *most* legitimate institutions of all those that are actually achievable. The fact that so many private initiatives start when official diplomacy fails hints that private bodies can sometimes reflect the popular will better than formal government.

It follows that even imperfect private governance should be acceptable where existing or reasonably foreseeable government institutions have shown themselves incapable of addressing some problem. This incapacity notably includes states that are "authoritarian and corrupt,"[34] captured by private clients,[35] or suffer from recurring deadlock.[36] That said, scholars stress that we should not lose sight of the opposite danger that private governance can equally transfer authority from legitimate states to unaccountable transnational bodies.[37]

The argument is more difficult in the broad middle case when government is functional but imperfect. Here the usual question is whether private bodies are inherently more efficient at certain tasks or can help their public counterparts realize synergies that neither could achieve alone.

Inherent Advantages. Scholars have argued that the New Self-Governance can potentially do many tasks better than states. First, government has only limited knowledge of what the people want. This suggests that New Self-Governance can sometimes provide information about the "public interest," thereby overcoming officials' "bounded rationality."[38] Second, self-regulation could develop better rules that accommodate business needs and individual situations while minimizing compliance costs and potentially damaging intrusions by outsiders.[39] Third, developing government consensus is expensive and there is also a "tendency [for] governmental action to become a smothering monolith." This means that it may be better for government to fix errors than take over the whole process.[40] Finally, government may lack monitoring resources, a problem that becomes particularly acute when activities are highly technical or take place outside its borders or in large, complex markets. This is the most common argument for domestic industrial standards.

The problem with these objections is that they are mainly anecdotal or ideological. The idea that conventional governments do such tasks badly is mostly an assertion. But in that case, it is hard to have any useful intuition about when to intervene. Timothy Lytton presents a more explicit and potentially useful theory of the problem by pointing out that government regulation budgets are set by politics whereas private regulation budgets are fee based and expand each time a new topic is regulated.[41] This means that official regulation can be oversupplied so that politicians and bureaucrats invest more than what the entire industry is worth to society. By comparison, society's investment in private regulation can never exceed what consumers are willing to pay for the regulated product.[42] On the other hand, government regulation can also be undersupplied, in which case officials may decide that it is better to wait for a crisis and even then deliberately limit the scope of new regulations.[43] By comparison, Lytton argues that private certification bodies tend to be more proactive in seeking out situations where they can improve matters and, not incidentally, charge more fees.[44]

Synergies. Private and public standards are particularly likely to generate synergies in international settings since democratic states (by definition) make no effort to represent noncitizens[45] so that independent decisions across multiple states are often inconsistent.[46] Introducing private governance in these situations plausibly reduces the danger of inconsistent regulation and facilitates formal harmonization later.

The downside of coexistence is that private standards can replace or override government outcomes. Indeed, our coffee, lumber, dolphin, and fish examples all began as attempts to evade a stalled diplomatic process. From this perspective, forcing the West's "embedded liberalism" onto global markets[47] seems painfully neocolonial. This objection can only be overcome by making private governance at least as democratic as the local governments it displaces. Efforts by 4C, FSC, and PEFC to include citizens from the developing world are a long step in this direction.

Finally, government could decide that it values private governance processes in their own right. This could reflect a pragmatic judgment that private groups are likely to adopt congenial or at least acceptable solutions on balance. But it might also reflect a more ideological commitment to democracy, bolstered by our shadow electorate arguments that the New Self-Governance is broadly representative.

7.3.5. Empirical Tests

In principle, at least, private bodies' responsiveness to shadow electors' values is observable. Most obviously, we can measure how much shadow electors know about private standards and how much corporate elites feel or act as if they are constrained by shadow electorates. This information could then be benchmarked against what voters in conventional democracies know and how much their representatives feel constrained by the need to win reelection. Scholars could also study the extent to which businesses' share prices and corporate behavior are affected by petitions and calls for boycotts. Finally, it would be instructive to examine the participation rate for different groups within the executive agencies (or possibly their iron triangles) compared to the players who appear and participate in private rulemaking. At least in our mutations case, private hearings seem to have been noticeably more inclusive and international than the corresponding official channels.

7.4. Deliberative Democracy

So far, we have stressed conventional democratic theories that let "the people" express themselves through voting. However, the difficulty of extending these ideas to private governance has encouraged many scholars to consider alternate democratic theories that elevate deliberation and participation from instrumental mechanisms to goals in their own right.[48] Similar claims are sometimes made for due process, rights of appeal, constitutionalism,[49] efficiency, expertise, effectiveness, due process, conformance with professional or scientific norms, and procedural values like consistency, proportionality, and so on.[50]

7.4.1. The Purpose of Deliberation

Many scholars argue that deliberation provides value in its own right.[51] From this standpoint, rules requiring notice and comment, structured adjudication by experts, transparency, "dialogue," and "scrutiny" are valuable simply because they stimulate "engagement" and increase the number of "participants."[52] Whether most humans share this view is ultimately an empirical issue. Even so, the idea that citizens would be content to deliberate without the slightest chance of affecting outcomes strains credulity.[53] In what follows, we assume that deliberation has some instrumental value in yielding outcomes that the wider society agrees with and is not just an end in itself.*

---

* At least formally, most scholars accept this view, acknowledging that new "form[s] of accountability" might be acceptable if they let "consumers express preferences ... they might otherwise express as citizens in the ballot box. Vanderburgh (2006/7) at p. 942; Scott et al. (2011) at p. 16 (arguing that attempts to achieve control of private power by transposing "norms and institutional structures" developed for formal governments needlessly limits the toolbox of potentially legitimate methods).

The clearest case for where deliberation is where we expect it to generate exactly the same outcomes that representative democracy would. We expect this to happen in three cases:

(a) The demos is sufficiently homogenous that everyone agrees on the same goals;
(b) The preferred solution simultaneously addresses all possible goals, so that normative choice is unnecessary; or
(c) Deliberation can change preferences and produce homogeneity even if it does not exist to begin with. This presupposes that normative preferences are either malleable or can be discovered through discourse.

The problem is what to do when none of these possibilities holds. The claim that broad participation can substitute for traditional liberal democracy is among "the most vexing normative implication" of New Self-Governance.[54] Despite this, we can imagine deliberation facilitating democratic outcomes in at least two respects.

Suppressing Self-Interest. Deliberation makes it harder for players to pursue self-interested outcomes without letting some audience detect and punish the fact. This argument is particularly strong in shadow electorate models where transparency helps consumers know when corporations have behaved badly and should be boycotted. However, one can imagine a similar dynamic where sanctions are not material but are instead limited to moral condemnation by some audience or, perhaps, the actor's own superego. In general, such theories require at least three necessary conditions: (a) an audience exists so that the deliberation does not just play to an empty theater, (b) the actors care what the audience thinks, and (c) this sanction at least sometimes outweighs the benefits of an unprincipled decision.

One of the most astonishing implications of deliberative democracy theory is that it authorizes small groups decide for everyone. This might not matter if we believed that the groups provided a reliable proxy for the rest of us. One hopeful datum is that business elites in Hollywood and Silicon Valley, say, often display surprisingly mainstream and even left-liberal views.[55] Similar observations almost certainly apply to science. At the same time, this coincidence is accidental and often fails in specific cases. U.S. courts and political philosophers have consistently argued that the bare possibility of abuse is unacceptable.[56]

Discovering Norms. Deliberative theory would make perfect sense if we believed that, given a long enough conversation, rational citizens would eventually reach a consensus on normative values in the same way that they agree on objective facts. Still, we should be skeptical. After all, philosophers have been pursuing this agenda since Aristotle's day. Even so, there could be special cases where people already agree on goals so that only the means are uncertain.[57] This is most likely to happen for topics like product standards and internet protocols where the affected groups are homogenous. Conversely, apparent agreement on normative matters could sometimes reflect ignorance. In that case, deliberation will lead to *less* agreement over time.* Probably the most important special case is where everyone agrees that "no standard" is the worst possible outcome. In these circumstances giving all relevant groups membership (and hence a veto) can force agreement around standards that everyone sees as improving.[58]

Traditional corporatist systems formalize such arguments by encouraging citizens to self-identify as part of larger groups, for example the four estates in eighteenth-century France or the employer/worker/government councils beloved of twentieth-century Fascists.[59] Suppose for example that members of FSC's various subchambers really did share identical views.[60] Then a sufficiently long deliberative debate within each sub-legislature would presumably produce answers that every member agreed with. More realistically, the fact that shorter debates produced a large if nonunanimous "consensus" could be taken as evidence that the minority would eventually be persuaded if the conversation continued. This, in turn, would help explain standards bodies' familiar insistence that consensus need not be unanimous: given that deliberation is expensive, real institutions will almost always find it cost-effective to truncate debate short of unanimity.† While there is always some risk that the majority is wrong, this is a transaction cost. At some point, it is better to stop deliberating and accept the possibility of error.

7.4.2. Institutions

Even if we believe that deliberative democracy is possible, success depends on detailed design choices. These typically include orderly debate procedures, public comment on proposed rules, and transparency.[61]

Multi-Chamber Models. We have argued that corporatist models are sensible when their legislative chambers accurately reflect the main divisions in society.

---

* This is more or less what happened in our fisheries example, where MSC agreed on democratic principles at the outset only to find that the supposedly technocratic choices that followed raised political issues they had overlooked.

† The fact that all real institutions have finite size tends to sweep this problem under the rug. Even apparent unanimity would disappear if the number of representatives was sufficiently expanded.

But it is not enough for chambers to be homogeneous; they must also be few in number. Otherwise, at least one body will almost always decide to veto the others. This explains why the number of chambers found in real organizations can usually be counted on the fingers of one hand. Here, it helps that private initiatives do not aspire to be general legislatures, since limiting decisions to specific topics facilitates "plausible claims to authority and common identity within a social or economic group."[62] Appeals that let outsiders challenge rules[63] similarly provide assurance that the corporatist architecture has not been gerrymandered to exclude important interests.

Beyond these basic choices, deliberative theories necessarily favor measures that let debate unfold in an orderly and predictable fashion. At this point, the recommendations of delibeative-democracy theory overlap those for representative democracy. These best practices include fair and open decisional procedures; the use of logic, evidence, and explicit reasoning- which make it is harder to frame arguments to reach predetermined results; and monitoring and learning from existing policies.[64]

7.5. Shortcuts

Our discussion so far shows just how daunting democratic legitimacy arguments can be. Not surprisingly, some scholars have looked for shortcuts.

7.5.1. Output Legitimacy

"Output legitimacy" argues that institutions are legitimate if they consistently produce desirable results.[65] However, "desirability" necessarily presupposes some (generally hidden) criterion for judging effectiveness and the common good.[66] In practice, the criterion can be almost anything, from generalized concepts of justice (e.g., Rawlsian conceptions of the "right")[67] to narrow policy issues (e.g., free trade)[68] to members' material self-interest.[69] The problem is that the legitimacy of these choices is almost always tacit, or assumed, or unexamined. For community-wide groups, the number of generally agreed upon principles is almost always limited.

7.5.2. Cognitive Legitimacy

"Cognitive legitimacy" rests on the "taken for grantedness" of familiar institutions, i.e., that it is literally unthinkable "for things to be otherwise."[70] A more principled justification, descending from Edmund

Burke, holds that evolved institutions often have virtues that we do not fully appreciate, so that change brings more harm than good. While the newness of private transnational institutions limits how much "taken for grantedness" is possible,[71] recurring calls to implement virtues like transparency import cognitive legitimacy from older and more familiar models. The question remains whether the mechanism's (possibly unknown) virtues will make sense in the new environment. If not, we should expect imported norms to be "fragile" once the concept is cut off from its original context.[72]

7.5.3. Semi-Empirical Theories

The Burkean worldview implies that humans can sometimes perceive legitimacy without being able to articulate their reasons. One way to formalize this instinct is to adopt a "popular" legitimacy approach that asks which arrangements people prefer in practice.[73] This deliberately blurs empirics and theory. On the one hand, the way people experience institutions can change which normative theories they find convincing. On the other, new institutions are more likely to be seen as legitimate if they are backed by coherent belief systems.[74] But in that case, it makes sense to treat the two as an undifferentiated whole.[75]

This opens the door to a kind of evolution in which legitimacy can grow over time.[76] First, pragmatic successes can accumulate so that "outcome legitimacy" increases. Later, legitimacy can transition from purely pragmatic grounds ("What have you done for me lately?") to deeper and more resilient normative principles. Finally, simple longevity can lead to a kind of Burkean legitimacy, which some scholars insist would be the most durable of all.[77] While the New Self-Governance is too new to have reached this last step,[78] the ad hoc importation of revered institutions like "transparency" and "consensus" from other settings probably introduces at least a measure of "taken for granted" legitimacy.[79]

7.5.4. Derivative Legitimacy

Scholars conventionally argue that traditional courts, legislatures, and executives lend their legitimacy to other bodies in the course of accomplishing their missions.[80] This "derivative legitimacy" approach is especially congenial to legal scholars, who often find it convenient to ask whether particular state organs are functioning "according to law" without triggering broader debates about whether the law itself is legitimate.

Delegation arguments are most obvious in traditional "shadow of hierarchy" situations where a (presumptively legitimate) state has deliberately established private governance to pursue its own ends. Here the main issue, as we will see in Chapter 8, is that citizens may not have authorized the state to delegate its power. However, our private-politics theory suggests a second ground. We have argued that the power that private organizations delegate to any particular group reflects bargaining against the latter's threat to leave the organization and petition other democratic channels if its demands are refused. However, this implies that power in private organizations is ultimately measured by the willingness of more traditional institutions (e.g. legislatures, judges, the press) to treat those claims as legitimate. In this sense, the influence that groups exercise *within* private bodies, far from being a free parameter, is almost entirely determined by the judgments of *outside* institutions that we already consider legitimate.

7.5.5. Other Theories

Democratic theories do not begin to exhaust the possible bases for legitimacy. These have changed constantly throughout history[81] and include source-based (e.g., religion, the people, the proletariat),[82] personal or charismatic,[83] and corporatist ideologies that root legitimacy in groups instead of individuals.[84] Many of these choices would make legitimacy analysis much easier than the democracy-based rationales discussed above. One possible use for deliberation is to identify and work out new norms for society to consider.[85]

7.6. Theories of Academic Legitimacy

Contemporary science-policy scholars overwhelmingly accept the postwar Republic of Science system in which government leaves detailed resource allocation to the scientists themselves.[86] This basic model satisfies conventional "derivative legitimacy" arguments in which a government consciously surrenders control, hoping to receive better science in return. Since the 1970s, the urge to inject outsiders – activists, social science faculty, and ordinary citizens – into community governance has grown in hopes of making the Republic of Science more sensitive to ethical and social issues without disrupting its core function of allocating resources. How this should be done is less clear. In particular, there are no natural criteria for deciding how many outsiders should participate, or who they should be, or how many votes (or vetoes) they should have. In the end, such discussions can only be evaluated in the context of some specific theory about how these outsiders could force change compared to the outcomes that scientists would adopt on their own. Echoing our commercial discussion, there are three possible mechanisms:*

> *Derivative legitimacy.* The most obvious way to give outsiders influence would be for the NIH to intervene in the event of a walkout. This would, of course, nothing more than the usual shadow of hierarchy dynamic. The question remains, however, whether threats to intervene are credible. First, the NIH has existed for decades. This strongly suggests that the agency may have been "captured," i.e. that regulators now identify more strongly with scientists than the public they are supposed to protect. But even if the NIH has not been captured, or is forced to act by public outrage and backlash, officials could still decide to do as little as possible on the theory that any intervention was likely to do more harm than good. Certainly academic scientists do not seem particularly worried about intervention in practice.†
>
> *Representative democracy.* We have emphasized that biotechnology and especially Big Pharma companies are sometimes answerable to shadow electorates of consumers. Where this happens, we expect private firms to press their academic partners to take the threat of walkouts seriously. One advantage compared to shadow of hierarchy models is that companies routinely hire science advisors to stay abreast of new academic discoveries. This probably means that they understand academic science much better than the NIH and, to that extent, are better placed to demand reforms.
>
> *Deliberative democracy.* The most hopeful scenario is that the goals of "responsible science" (e.g., new knowledge, higher living standards, safety and security) are so widely agreed upon that any remaining debate can be limited to an objective discussion of means. In this view, the public might reasonably think it would reach the same results as scientists if it understood the debate. The role of outside observers

---

* For concreteness, we focus on the case of NIH and academic biology which has historically dominated the discussion.

† For example, the FBI stressed in a 2009 meeting "that scientists are responsible for how science gets used" and that biologists "must display that they can do the research safely and securely in partnership with the FBI, or face potential legislation." Grushkin (2010) (quoting Special Agent Ed You). Despite this very explicit threat, the community never even tried to self-regulate.

would then be mostly to confirm that scientists were conscientiously pursuing these issues despite occasional self-interest. Rules that provide for orderly political debates would similarly demonstrate that such self-interest as did exist would be promptly exposed, confirming that the process as a whole was trustworthy.

8

Law

The last chapter argued that our shadow electorate model provides a natural extension of traditional legitimacy arguments in which popular elections reflect "the will of the people." This has immediate, practical consequences. Judges in common law systems are constantly asked to rethink existing rules, identify incoherencies, and fix them. This almost always means returning to the law's original intent along with any "latent" issues that the drafters may have overlooked the first time. We argue that our shadow electorate model – where market prices are cartelized but quality competition is still strong enough to empower strong shadow electorates – presents just such an issue while offering, at least potentially, a far more principled framework for deciding when private standards are acceptable.

We emphasize that, far from being a mere theoretical exercise, the need for improved legal rules is intensely practical. The existing law of private power is so unclear that most real-world initiatives spend enormous resources funding legal research,[1] imposing awkward, ad hoc rules about what members can say to each other;[2] and generally debating which kinds of collective action are and are not lawful. Given the potential ferocity of antitrust sanctions, it is reasonable to assume that many ventures are deterred entirely. This situation is especially frustrating for U.S. government agencies that see self-governance as a policy tool and try to reassure industry they will not prosecute innocent violations.[3] The trouble, of course, is that none of these statements is binding. This leaves companies wondering what will happen if they misunderstand regulators' assurances, or the political winds shift, or the current administration is voted out of office. In any case, the Sherman Act authorizes private lawsuits. So long as the law authorizes intervention, no one should expect antitrust courts to turn a blind eye.[4] The only honest and coherent remedy is to reform the law itself.

This chapter reviews the law of private self-governance. Section 8.1 presents basic common law and constitutional doctrines that set minimum requirements for private self-governance, usually watered-down versions of practices inherited from government agencies or the judiciary itself. Section 8.2 introduces the U.S. Sherman Act in its original guise as a tool for limiting private political power. This line of cases is still binding today. Section 8.3 extends this discussion to include modern approaches that recast the Sherman Act in terms of economic efficiency. We particularly focus on how self-governance initiatives can structure themselves to minimize compliance issues.*

8.1. Common Law and Constitutional Limits

Limits on private power are rooted in the U.S. Constitution and judge-made "common law" dating back to the Magna Carta. These rules set minimum standards that no policymaker can change.

Common Law. Common law courts analyze private power using the same legal theories that govern interactions between individuals. This means, for example, that litigants who belong to a particular self-governance body can bring breach-of-contract actions to enforce express or reasonably implied promises. Even without such promises, organizations must honor everyday expectations of fairness under fiduciary duty theories.[5] Similarly, nonmembers can sue private bodies under various tort law theories, including defamation and interference with economic advantage.[6]

Courts often claim that they are prepared to overrule private rules on substantive grounds.[7] However, this seldom happens in practice absent direct conflicts with statute or very clearly stated public policies.[8] The reason, almost certainly, is that unelected judges are reluctant to legislate for others. Courts have, however, shown a modest appetite for striking down private rules that were imposed unfairly or violate basic democratic norms. This includes rules adopted without adequate notice, opportunity to comment and debate, or majority vote.[9] Courts are similarly willing to overturn enforcement where the accused has not received a good-faith hearing, including both reasonable notice and an opportunity to prepare.[10] While these requirements are plainly modeled on courts and the executive branch, acceptable private procedures can be much less formal. This is especially

---

* Readers seeking additional legal details can find them in Maurer (2014).

true where industry goals do not impinge on the public interest or the accused's right to earn a living;[11] the private body possesses "specialized competence";[12] there is no visible conflict of interest;[13] or higher standards would make self-governance impractical.[14] However implicitly, this deference implies that courts see private governance as valuable and legitimate."[15]

Constitutional Issues. The U.S. Constitution protects citizens when private associations perform "governmental functions by the sufferance of the state." In practice, this category is usually confined to semiofficial bodies like state universities or bar associations.[16] Additionally, the Constitution's "nondelegation doctrine" prevents some government functions from being outsourced at all. While no court has yet considered the issue,[17] the U.S. government has said that its agencies cannot let private bodies exercise military, diplomatic, civil, or criminal functions, manage government contracts, take actions that significantly affect the life, liberty, or property of private persons, or control the acquisition, disposition, and use of federal property.[18]

Impact. In practice, these common law and constitutional limits on self-governance have had relatively little impact. This is because large modern initiatives like our timber, fisheries, and coffee examples almost always choose much more elaborate rules than U.S. law requires. Common law and constitutional rights do, however, help ensure that these organizations abide by their own procedures. Smaller bodies without formal rules (e.g., artificial DNA) provide a closer case, although even here courts usually uphold almost any procedure that is administered fairly.

The question remains what judges should do when private bodies embrace safeguards that look very different from conventional government. To date, no court seems to have considered this issue. At least in principle, however, courts should be willing to tolerate weaker procedural rules where anchor firms have established alternative bases for legitimacy by sharing power with outside interests and/or engaging rival standards in vigorous external politics.[19]

8.2. Antitrust

Judges often discover new purposes in old laws. America's Sherman Act was originally conceived as a *political* response to private power. Similar concerns helped launch competitions policy in Europe and Japan after World War II,* most notably through German ordoliberal

---

* European Union competitions policy grew out of the disastrous role that monopolists played in the rise of Hitler and was further shaped by German ordoliberal traditions that

tradition.* Even so, phrases like "restraint of trade" or "monopolization" *sound* economic. Over time, this encouraged judges to understand the statute in the elegant language of microeconomics. The result, a century later, is that the Sherman Act has two faces. Subsection 8.2.1 examines competitions policy from its original political purpose. Subsection 8.2.2 examines today's dominant economic-efficiency interpretations, stressing practical steps that self-governance bodies can take to avoid liability.

8.2.1. Sherman Act Originalism: Limiting Private Power

We saw in the last section that common law judges typically let litigants enforce contract rights against private governance bodies. But one might equally ask why governance bodies do not ask courts to enforce their rules against members. The reason, usually, is that the Sherman Act limits agreements across firms. This makes sense since, as Ian Maitland observes, "The powers to prevent trade abuses are the same powers that would be needed to restrain trade."[20]

Microeconomics barely existed when Congress passed the act in 1890.† As Richard Hofstadter has stressed, the act's original goals were instead primarily concerned with the *political* argument that private wealth would capture government and produce a plutocracy.[21] This was understandable in the late nineteenth century, when giant corporations were becoming prominent. In the words of the act's sponsor, Sen. Senator John Sherman, "some corporations were big enough to dominate state governments, and if they should combine among themselves, they might come to dominate the federal government as well."[22]

But political power is hard to measure: just when, exactly, does size become a threat? Looking back, the genius of Sherman's approach lay in translating a political idea into the language of markets. "If we will

---

stressed the need to protect individuals against both state intervention and private abuse. See, e.g., Anchustegui (2015); Gerber (1994). American occupation authorities similarly introduced antitrust law to Japan because of the role that the country's Zaibatsu conglomerates had played in its slide to war. Bisson (1954).

* This was particularly true among the first generation of ordoliberalism scholars, notably including Freiburg University's Walter Eucken. Wikipedia (n.d.(l)). The Freiburg School often referred to the argument as "vollständige Konkurrenz" or "complete competition." Florian Möslein (personal communication).

† Modern microeconomic theory is conventionally dated from publication of Alfred Marshall's *Principles of Economics* in 1890, the same year that the U.S. Sherman Act was enacted. Hofstadter's thesis should not be overstated: The Sherman Act was never *entirely* political, but contained consumer protection elements from the outset. We address the latter in Section 8.3 below.

not endure a king as a political power," he thundered, neither should we "endure a king over the production, transportation, and sale of any of the necessaries of life."[23] This rhetorical equivalence turned out to be intensely practical: compared to political power, economic power is measurable and precise. Furthermore, it is subject to a variety of shorthand empirical tests, like whether the dominant firm has "the power to prevent competition," "fix the price of any commodity,"[24] or "command[] the price of labor."[25] Early Supreme Court cases further refined this approach by interpreting the statute as an attempt to stop "any one commodity" from falling "within the sole power and subject to the sole will of one powerful combination of capital."[26]

The Supreme Court's *Standard Oil* decision[27] completed the process of identifying economic power with the newly completed framework of classical economics.[28] Despite this, the Sherman Act's original political purpose remained, as lawyers say, "good law." Indeed, it experienced a renaissance of sorts in the 1930s when Fascist political models claimed that private groups should assume government-like powers over the economy.[29] This was too much for the U.S. Supreme Court, which struck down President Franklin D. Roosevelt's own experiment with worker-management councils in 1935.[30] That case, however, was limited to Congress's power to delegate core government functions to the private sector. This left open the question of whether private bodies could seize such power without Congress's help. The question became urgent when an industry group organized the Fashion Originators' Guild of America (FOGA) to enact and enforce "ethics" rules for selling women's clothing.

The U.S. government's Sherman Act prosecution against FOGA reached the Supreme Court in 1941.[31] While the Court was willing to strike down FOGA on economic efficiency grounds,[32] it also went out of its way to recall the Sherman Act's older political purpose. This led to a much broader holding that FOGA was likewise objectionable as "an extra-governmental agency, which prescribes rules for the regulation and restraint of interstate commerce, and provides extra-judicial tribunals for determination and punishment of violations, and thus 'trenches upon the power of the national legislature' and violates the statute."[33]

The *FOGA* decision was a clear repudiation of private power. But it was also incoherent. Read literally, the Court seemed to have banned private self-governance altogether, particularly when private groups claim the power to punish.[34] Yet, that reading could not be true: if the Court had meant to abolish thousands of private standards across the American economy, it surely would have said so. This left lawyers and judges struggling to articulate some sensible dividing line. The question boiled down to deciding when "the external impact of the standard setting" becomes so severe that "no private group is a safe repository for such quasi-governmental power."[35] *FOGA*'s progeny did little to address this question. Instead, judges tried to distinguish "coercive" rules from those that were said to be "truly voluntary," "merely persuasive," or limited to social "naming and shaming" sanctions.[36] The problem with these formulations was that they were never grounded in any obvious policy logic. This meant that the proposed distinctions were mostly semantic and circular – pressure is fine, the cases say, but only until it becomes "coercion." In practice, the language seldom mattered since hardly any practices were struck down. Most scholars take this as a hint that *FOGA* is a dead letter.[37]

Not surprisingly, firms see things differently. Given that antitrust prosecution is a calamity, why run even a small risk? The good news, following our commercial examples, is that firms often pursue self-governance anyway. The bad news is that there is no way of knowing how many other industries have been deterred or else limited themselves to the short list of methods ("naming and shaming," expelling violators from trade groups) that courts have blessed.[38] This leaves the New Self-Governance's most potent weapon – adopting standards that reduce dissenters' profits – in limbo.

Doing Better. The difficulty in defining "coercion" is, of course, identical to the old debate over "power." Seven years after *FOGA*, Justice William O. Douglas gave the same answer that John Sherman had by proposing economic power as a stand-in for private political leverage. Writing for the dissent, his words are not binding as precedent. Despite this, they are widely respected among antitrust lawyers. Granting "a handful of men" the power to control steel prices, Douglas wrote, would potentially allow them to wreck the economy. From this standpoint the "problem of bigness" was not just economic but also "a social menace":

> That power ... can be benign or it can be dangerous. The philosophy of the Sherman Act is that it should not exist. For all power tends to develop into a government in itself. Power that controls the economy should be in the hands of elected representatives of the people, not in the hands of an industrial oligarchy. Industrial power should be decentralized. It should be scattered into many hands so that the fortunes of the people will not be dependent on the whim or caprice, the political prejudices, the emotional stability of a few self-appointed men. ... That is the philosophy and the command of the Sherman Act. It is founded on a theory of hostility to the concentration in private hands of power so great that only a government of the people should have it.*

---

* *United States v. Columbia Steel Co.*, 334 U.S. 495, 535–6 (1948) (Douglas, J., dissenting). Judge Learned Hand made much the same point in the First Circuit's *Alcoa* decision: "We

Seventy years later, the Sherman/Douglas argument remains the clearest suggestion for a way forward. Its great advantage lies in reducing a nebulous judgment about political power ("coercion") into a well-defined economic inquiry ("controlling steel prices"). Its main drawback, like all bright-line tests, is that it does not even try to make reasoned distinctions about what is "unacceptable" in close cases. The power to wreck the economy seems clear enough. But what should courts do when private power is tiny, or can only be exercised for brief periods, or within some narrow price band? At some point, judges will say these cases are too insignificant to bother with. They will almost always be right without being able to say why.

Extending Doctrine. This book has argued that quality competition can impose shadow electorates that limit private power in the same way that ordinary price competition does. But the Sherman/Douglas test overlooks this possibility. Here, the most straightforward reform is to extend the Sherman Act to say strong *quality* competition *also* makes private power acceptable, at least when it rises to such levels that executives feel compelled to anticipate and respect consumers' policy choices. Subsidiary factors would then include whether anchor firms have taken steps to share power with interested groups,* adopted institutions that either promote or have actually achieved transparency, and/or are strongly constrained by competition with rival standards.

Needless to say, these suggestions leave many details for judges to fill in later. This would presumably be done in the usual common law way by repeatedly applying (and if necessary modifying) the proposed rules in light of experience. In the meantime, a Sherman Act analysis extended to include quality competition and shadow electorates would provide a natural way to move beyond today's standardless inquiries.

---

have been speaking only of the economic reasons which forbid monopoly; but, as we have already implied, there are others, based upon the belief that great industrial consolidations are inherently undesirable, regardless of their economic results. In the debates in Congress Senator Sherman himself ... showed that among the purposes of Congress in 1890 was a desire to put an end to great aggregations of capital because of the helplessness of the individual before them." *United States v. Aluminum Co. of America*, 148 F.2d 416, 428–9 (2d Cir. 1945).

* This would itself require line drawing by the courts. See, e.g. Leibowitz (2005) ("Now, I don't think that board full of industry competitors becomes diverse with the addition of a single stray 'public' member ... but it is hard to say in the abstract when the composition of a state board is sufficiently diverse that the board does not have to be supervised to have recourse to the state action defense").

8.3. The Sherman Act Reimagined: Economic-Efficiency Arguments

The Sherman Act injected previously undefined terms like "restraint of trade" and "monopolization" into the law. The Supreme Court's landmark *Standard Oil* decision told judges to interpret these concepts in terms of economic theory. The following sections introduce the modern framework and its applications to industry self-governance.

8.3.1. Overview

Sherman Act § 1 declares "[e]very contract, combination ... or conspiracy, in restraint of trade ... illegal."[39] On the other hand, no one argues that Congress wanted to ban contracts indiscriminately. The better question is therefore which agreements remain permissible. The statute itself is not much help. The basic difficulty is that *all* contracts restrain trade: if A promises to sell eggs to B tomorrow, A immediately loses the right to sell those same eggs to C this afternoon. The *Standard Oil* case (1911) cut this Gordian knot by inventing a "Rule of Reason" that courts would only strike down agreements whose "pro-competitive features" are outweighed by "anticompetitive effects." But this is still not sufficient since the links between "features" and "effects" are often obscure. Modern paraphrases avoid the problem by saying that the antitrust laws condemn agreements that "raise prices or reduce output."[40]

Applying this framework to self-governance raises two distinct issues. First, any attempt to adjust market outcomes to include previously ignored social values ("externalities") implicitly or explicitly starts by assigning a price to them. Whether the Sherman Act allows private groups to do this remains unclear. Second, even assuming this power, private interventions must be designed to address the social problems at issue without fixing prices and output more generally. Getting these rules right turns out to be surprisingly subtle.

8.3.2. Pricing Externalities

We have already said that private agreements should not raise prices or reduce output. But this immediately runs into the objection that practically all meaningful self-governance does precisely that. The usual answer is to say that this is only an accounting illusion. Granted that measures that reduce dolphin mortality *seem* to raise tuna prices, this is only because the value of dolphin lives has been artificially (and incorrectly) set to zero. It follows that any reformed calculation must start by attaching a more accurate price tag to dolphin mortality.

The rub, for judges, is that the revised price tag includes the dollar value of "*both* monetary *and* ethical"[41] goals. Since the latter have no established market, any judgments are unavoidably arbitrary. Before the 1970s, courts usually avoided this issue by arguing that the antitrust laws did not apply so long as industry or professional bodies served some facially noncommercial purpose.[42] Commercial bodies could likewise impose restrictions so long as they appeared to serve some legitimate, nonmarket goal.[43]

The trouble with these doctrines was that, notoriously, professional and industry groups often use "ethical" rules to suppress competition. The Supreme Court's *Goldfarb*[44] decision acknowledged this reality in 1975. Even so, the Court hinted that professions could still adopt restrictions "which could properly be viewed as a violation of the Sherman Act in another context" for valid, noncommercial ends.[45] Indeed, some justices suggested that private standards could still "take account of benefits other than increased competition,"[46] provided that the "restraint is not unreasonable in its effect on competition."[47] The problem, as always, was the lack of "objective benchmarks"[48] for weighing noneconomic factors.[49] The issue finally came to a head in *NCAA v. Board of Regents of the University of Oklahoma*,[50] where the Court upheld rules regulating student athletes' "amateur" status. But while the Court admitted that the rules served a noneconomic, educational purpose, it claimed to decide the case on the purely commercial ground that amateurism made college football more interesting.[51] Most commentators,[52] courts,[53] and lawyers[54] have taken this as a hint that noneconomic goals are off-limits, despite the fact that many lower courts continue to invoke ethical goals as well.[55] There is also some suggestion that noncommercial rules can be justified on the backdoor principle that companies have an interest in forestalling moral outrage that might destroy their businesses.[56]

In some ways, the impasse resembles our earlier *FOGA* discussion: while courts claim to ignore noncommercial values, they nevertheless uphold them. The basic difficulty is that literal implementation of the rule automatically sets externality prices to zero. This seems inconsistent with the usual antitrust consensus that self-regulation can be used to overcome market imperfections.[57] From this perspective, letting industry attach a price tag to social concerns can only improve matters.* Some

---

* In theory, industry could overregulate itself so that society received net negative value. This outcome is contrary to the usual assumption that firms are strongly biased toward minimal self-governance that keeps their costs as low as possible.

European scholars already argue that competitions law should regulate cartelization for public purposes instead of banning it.[58] In what follows, we assume that agreements that encourage more spending to address social problems are legal so long as they do not raise prices and restrict output more generally.[59]

8.3.3. Rule of Reason

Lawyers, notoriously, like broad, predictable rules. So it should not be surprising that early antitrust practitioners looked forward to a day when *Standard Oil*'s open-ended Rule of Reason standard would give way to detailed, hard-and-fast ("*per se*") rules for the great majority of cases. But this proved harder than expected: by the 1930s, the Supreme Court had largely given up on this agenda.* The result today is that most antitrust cases are considered on their individual facts, with each Rule of Reason analysis proceeding from economic arguments at least as much as legal principles. Like most lawyers and competitions policy officials, we follow this approach in what follows.

Consider, then, how different types of self-regulation translate into changes in price or output. Almost all real initiatives fall into one of three categories:

> Case 1. *Specifying goals.* The organization can set standards that define, for example, what level of dolphin mortality or bacterial level in food is acceptable. So long as companies satisfy these "metrics," they can adopt any methods they want.
>
> Case 2. *Specifying methods.* The organization can set standards that ignore outputs but require members to adopt defined methods and equipment. This is especially attractive in cases (e.g., forestry, fishing) where the available metrics are unreliable or only distantly related to goals.
>
> Case 3. *Providing solutions.* The organization can set standards that require specific solutions and charge members for using them. In principle, this lets firms share costs, access economies of scale, and pool their best ideas to develop solutions in a cost-effective way.

From an antitrust perspective, Case 1 is by far the most straightforward. Changed operations will indeed impact prices and output through cost. But this will still be an improvement if the externality has been priced

---

* The late Suzanne Scotchmer and I have argued that the Supreme Court may have given up prematurely. Maurer and Scotchmer (2007).

correctly. Moreover, price increases are almost certain to be honest since firms continue to compete: this ensures that higher prices really do go to compliance instead of inflating profits. Probably the main concern in these cases is whether companies can design performance goals that exclude would-be competitors. This is most likely to happen where firms use different manufacturing technologies, so that compliance is more expensive for some firms than others. Alternatively, firms could use standards to keep profits too low to attract new firms that would drive prices even lower.* Fortunately, these strategies usually require fairly special facts. While courts should be sensitive to the possibility, they are unlikely to encounter them often.

The analysis for Case 2 is similar, except that companies must now adopt specific methods and technologies. This almost always increases the danger to competition by favoring some manufacturers' technologies compared to others. Worse, it suppresses competition to find more capable or lower-cost methods. This may or may not be justified to the extent that measuring outcomes like environmental impact or terrorism risk turns out to be hard.

Case 3 is by far the most difficult. So long as royalties are small and plausibly related to R&D costs, courts can simply decide to ignore them. Otherwise, the proper antitrust treatment will usually depend on how the royalties are structured. Royalties that scale with the number of units sold can be used to force up prices even when the resulting profits are distributed back to members.† This creates the obvious possibility of "sham agreements" that license technologies not because they are valuable but because they provide a handy way to organize cartels. This is at least consistent with persistent rumors in our artificial-DNA example that some firms saw high royalties for screening software as a way to drive smaller firms from the industry. Standards organizations can dispel most of these suspicions by sharing R&D costs through up-front or one-time payments, provided that these are not large enough to deter new firms from entering the industry.[60]

---

* Similarly, von Engelhardt and Maurer (2012) show that standards that inflate industry costs can sometimes stabilize high monopoly prices by making it unprofitable for new firms to enter the market.

† The assertion follows from freshman economics: given that firms set prices equal to marginal cost, adding a royalty to each unit produced immediately raises prices. The fact that royalties are later rebated does not matter so long as rebates are proportional to industry-wide sales rather than each member's purchases. For a rigorous discussion, see Maurer and Scotchmer (2014) at pp. 314–16. One particularly elegant way to stop shared facilities from charging monopoly prices is to adopt bylaws that prevent them from paying out rebates to members. This prevents any monopoly profit from reaching members, so that the facility has no incentive to charge monopoly prices in the first place. Id.

Sham Agreements. We have already noted how rules nominally designed to pursue social issues can raise prices. But this is not the only danger. Most real-world cartels fail because it is hard to detect and punish members who break promises. This sets up the usual instability in which members who ignore their cartel commitments take business from those who honor them, so that each member finds it more profitable to break their promises until the agreement unravels. One way to stop this is to conclude seemingly innocent "sham" agreements that authorize acts (e.g., monitoring) that make it easier for members to detect and punish cheating that would otherwise destabilize the cartel.

This is reasonable so far as it goes. But the evidence for illicit agreements is almost always inferential, encouraging judges to find conspiracies where none exist. This implies that antitrust risk can never be reduced to zero, and to that extent is a drag on otherwise useful self-governance. One partial solution is to exercise more deference where anchor firms plausibly delegate power to groups that can be reliably counted on to detect and expose cartelization.[61]

8.3.4. Monopolization

We have so far concentrated on contracts in restraint of trade. However, very dominant companies can impose standards without any agreement at all.* Sherman Act § 2 addresses this possibility by banning unilateral acts that protect an existing monopoly or establish a new one. This analysis almost always returns to our earlier arguments that incumbents can sometimes manipulate self-governance so that new entry becomes less profitable. The difference in practice is that § 2 is limited to a relatively small number of well-defined triggering acts (e.g., "predatory pricing" or "tie-ins").† This leads to the ironic result that monopolists who impose private standards unilaterally are less likely to encounter legal challenges than they would in markets where self-governance requires agreement by multiple firms.

8.3.5. Procedural Safeguards

Congress has repeatedly authorized industry groups to pass rules that would normally violate the Sherman Act. The Supreme Court's *Silver*

---

* This arguably includes our StarKist example in Chapter 2.
† *Verizon Communications v. Law Offices of Curtis V. Trinko, LLP* 540 U.S. 398 (2004). European courts appear noticeably more willing to consider "essential facilities" claims. See, generally, Maurer and Scotchmer (2014).

(1963) decision limits this delegation by specifying that such rules are only valid to the extent that they serve "legitimate self-regulative ends."[62] The Court added that organizations must also take steps to reassure outsiders that the claimed "ends" are genuine and not just "sham" excuses to justify an illicit cartel. This includes adopting sufficient procedures so that outsiders can confirm that self-regulation addresses legitimate problems,[63] identify "unsupportable accusations" for which no evidence exists, detect evidence of ulterior motives, and generate a record for antitrust courts to review. Lower courts have since extended *Silver* beyond congressional delegations to include cases where self-governance is unavoidable because markets would not exist otherwise.[64]

In practice, *Silver*'s good faith, notice and comment, and hearings requirements are nearly indistinguishable from the constitutional and common law rules discussed in Section 8.1. However, it is at least arguably more intrusive where rules touch on antitrust's core concern with preserving competition.[65] The requirements are also subject to balancing. Thus, *Silver*'s requirements cannot be so burdensome that they "undermine" the association's "very authority ... to impose sanctions."[66] Conversely, courts can demand more elaborate procedures when the private association wields "tremendous economic power"[67] or controls the right to pursue a business.[68] Once again, these balancing approaches imply that private standards are legitimate and deserve deference even when the court might have preferred some other rule on substantive grounds.[69]

9

Policy and Practice

The New Self-Governance is here to stay. It follows that responsible officials have an obligation to interact with and sometimes manage it. But how? Should they encourage collective action? Press private bodies to embrace specific outcomes? And if so, what methods should they use?

We begin with commercial science. Section 9.1 starts with the basic question of when officials should promote private governance in the first place. Section 9.2 then asks what policy levers can be used to promote and influence private initiatives. Section 9.3 examines how government should manage the inevitable conflict where private and public regulations intersect. Section 9.4 extends these analyses to noncommercial science. Finally, Section 9.5 asks what public policy scholars can do to support these endeavors.

9.1. Strategy: The Decision to Intervene

We saw in Chapter 2 that governments have often supported private governance for ad hoc reasons, usually to bypass stalled treaty negotiations (e.g., fisheries) or legal prohibitions (e.g., lumber) against direct government action. In the long run, however, this sort of opportunism is disabling: Society will never get full value from private governance until officials can systematically examine every project to see which ones should or should not be encouraged. The balance of this section examines the possible advantages and feasibility issues that officials should consider in deciding when the benefits of private governance outweigh the cost of intervention.

9.1.1. Benefits (A): Pragmatic Advantages

The simplest justification for private governance is that firms possess identifiable resources that let government achieve more than it could on its own.[1] These typically include:

Physical and human assets. Private manpower and capabilities in many industries (e.g., biotechnology, computer security) routinely exceed the government's. Furthermore, many security tasks fit naturally into existing business operations, implying that the marginal cost of intervention is low. Resource constraints are particularly important for governments in the developing world that lack either the information or the competence to regulate.[2]

Policy and rule making. Government officials have limited resources to study problems and promulgate rules. This implies that most problems will go unregulated unless the private sector steps in.[3] U.S. officials have long recognized that private organizations often possess more information and can produce better standards.[4]

Information assets. Firms often possess uniquely valuable information about their internal operations, competitors' activities,[5] and the economic and technological feasibility of standards.[6] NGOs may similarly know more about social needs and the private sector's conduct on the ground.[7] Finally, private standards can generate new knowledge through research and experimenting with new standards to test their feasibility, benefits, and public acceptance.[8]

Monitoring and enforcement. Governments have limited monitoring and enforcement resources. This is especially true in complex markets,[9] beyond national borders,[10] or inside developing countries.[11] By comparison, many private firms have constructed elaborate institutions to monitor and manage their global supply chains. These commonly exceed host nations' capacities.[12]

9.1.2. Benefits (B): Systemic Goals

So far we have assumed that government has limited resources but is otherwise perfect. This is plainly untrue, suggesting that private governance is at least sometimes more efficient.

Institutional Frictions. Government institutions are shaped by long histories that encode past conflicts and compromises,[13] coalition politics that couple otherwise unrelated issues, and bureaucracies that harbor careerist and ideological ambitions. These frictions make formal regulation expensive and slow.[14] This penalty is further compounded where interventions are economically or scientifically uncertain* so that solutions require multiple trial-and-error experiments.[15] In extreme cases, conventional government can be degraded to the point where it becomes ineffective,[16] captured or corrupt,[17] or entirely absent. When this happens, private sector organizations – for all their imperfections – become credible alternatives.[18] The most obvious practical test is whether conventional government has tried and failed, sometimes repeatedly, to address particular policy problems."[19]

Scale. Policymakers should ideally balance the economies of scale from developing and administering a single standard against the need to accommodate local conditions and political preferences. The problem for governments is that scale economies almost always stops at national borders, which are themselves the arbitrary product of forgotten wars. Pushing beyond this limit requires governments to coordinate, implying a correspondingly steep increase in cost and delay.[20] Worse, competition between states – for example to attract business – can sometimes set off a destructive race to enact ever looser rules.[21] By comparison, private standards are coextensive with markets, whose boundaries continually adjust to realize emerging scale economies. This often encourages private bodies to areas up to and including the entire planet.[22]

Political Discovery. Private standards provide a unique testbed for discovering what the public actually wants,[23] providing occasions for groups to negotiate new coalitions[24] and demonstrating those alliances in miniature within private governance bodies that include hundreds and sometimes thousands of members. This gives public politicians crucial information about whether full-scale versions can be implemented on a national scale.

Fact Discovery. Private governance often expands traditional fact-finding methods based on hearing and debate to include external competition.†
This method seems particularly suited to science and economic judgments that require professional training or are otherwise too complex for regulators to verify from submitted documents. Despite persuasive anecdotal evidence,‡ we would like to know more systematically whether this is true.

---

* Our examples show that even industry insiders often know surprisingly little about, for example, the extent to which improved coffee standards would increase supplier profits or the best way to screen artificial-DNA orders.

† We have already emphasized that external politics presupposes an audience of shadow electors.

‡ Big companies in our artificial-DNA example routinely claimed that strong U.S. government regulations were unaffordable and would drive employers overseas. Conventional

Enforcement. Regulation ignores and sometimes tries to overrule market signals. This often leads to situations where firms that cheat become more profitable. This, in turn, can create a political dynamic that encourages foreign governments to refuse or slow-walk enforcement.[25] At least potentially, the New Self-Governance avoids these problems by flipping incentives so that compliance makes firms *more* profitable than they would be otherwise.[26]

Democracy. Government's monopoly of physical force lets bureaucrats tell people what to do. This can be a seductive alternative to endlessly redrafting documents to take account of additional opinions, local circumstances, and new knowledge and ways of doing things.[27] Private organizations are more fragile: leaders in our fisheries and lumber examples plainly understood that insisting on an unpopular standard could destroy their organizations. This explains the common observation that private standards are "accessible to a much wider group of players and may be nimbler and more portable."[28]

9.1.3. Costs and Downsides

Private standards also have drawbacks. Probably the most common objection is that they have less bite: bluntly, they cannot send dissenters to prison. The better question, however, is whether they have *sufficient* bite to accomplish their purpose. For firms, the prospect of negative profits is already a kind of death penalty. Furthermore, power is multidimensional: whatever weakness private standards may display in sanctioning individuals is often offset by their ability to reach much larger populations. Changing the procurement practices of a Walmart or Home Depot has "major global social and environmental impacts, comparable to if not greater than that of many national regulations."[29]

Private governance also requires a learning curve. To succeed, officials need to learn new ways of doing business that owe more to Silicon Valley than Washington, DC. Despite this, academics and policymakers almost always talk about organizing meetings to "engage" industry "stakeholders" in creating agreed standards. However, these formulations recast private politics in exactly the same terms that government uses to describe itself. This necessarily ignores the economic basis of private

---

hearings generated contradictory testimony but were unable to settle the issue. By comparison, the fact that relatively high-cost European companies had voluntarily adopted strong private standards immediately showed that the supposed burdens were modest.

power and, with it, some of private power's most attractive features. In particular, private standards may require much less consensus than officials imagine. To cite an admittedly extreme example, Bill Gates never asked stakeholders whether Windows should be the world's dominant software standard. Instead, the number of Windows adherents simply grew to the point where the market decided – and those who disagreed went out of business.

Finally, officials who promote private solutions will almost always forfeit a certain measure of control. Traditional shadow of hierarchy analysis justifies this trade as a quicker path to government's own goals. The difference in the New Self-Governance is that the goals will often be set by others. This should be acceptable where officials believe that private bodies produce the same outcomes that they themselves would reach on average. This is a fortiori true where government cannot achieve its goals directly or would have to overcome opposition from other nation states. The more distant prospect is that officials could eventually decide to defer to private political outcomes as democratically legitimate even when they themselves would have preferred a different outcome.

Government officials are bound to find private governance unfamiliar and even counterintuitive at least at first. The good news is that anything they learn can be reused. The bad news is that officials who avoid experimentation may never know what is and is not possible. The commonsense advice is that officials should seek out modest experiments and then try again if the results are encouraging.

9.1.4. Feasibility

Finally, policymakers must judge when private governance is feasible. This includes confirming that proposed projects enjoy reasonable political support in the community, the proposed strategy is capable of achieving its goals, and those goals are themselves lawful and legitimate.

Power. Policymakers should insist that would-be organizers are able to articulate at least one strategy where private power is both possible and politically achievable. Crucially, we have seen that organizers seldom know very much about what has or has not worked in the past. In these situations, officials will often be able to evaluate the proposed scheme better than the community itself.

This will plainly require clear thinking. The conceptual starting point, as we argued in Section 3.2, will be to decide whether the proposed standard is meant to (a) empower individual choice, or else (b) spark collective action and coerce dissenters. Depending on the answer, officials will then face two very different feasibility judgments: Where the goal depends on empowering choice, the key questions will usually turn on whether consumers really do want higher standards in the first place. Absent strong evidence, officials should worry that private standards will turn out to be pointless because most consumers either prefer low quality, or else are evenly balanced so that consumers who prefer high standards simply trade places with those who do not. By comparison, projects that require collective action imply an entirely different analysis. Here, officials will normally start by confirming that the targeted industry fits Chapter 3's basic New Self-Governance template in which (a) large manufacturers and retailers dominate concentrated downstream markets, while (b) upstream markets are populated by relatively small suppliers with large fixed costs. Once these threshold facts are established, officials should then look for additional evidence that preferred supplier agreements are widely used and/or intelligent-actor threats are present. These will normally provide strong "plus factors" that private governance is feasible.*

Politics. Assuming that private power is possible, officials will still want to know whether members are politically prepared to accept useful amounts of self-regulation. The most obvious course is to poll community members. However, our examples suggest that officials will often encounter situations where members have yet to form definite opinions or, if they have, are reluctant to disclose them.† At this point, officials will have to fall back on consulting self-announced players and organizers. This will often be a sensible proxy: if organizers have decided to invest their own time, that is already a signal that others can be persuaded.

Beyond this, officials should pay close attention to would-be leaders' personal traits. Most of our private governance examples would never have happened without strongly committed personalities like Markus Fischer (artificial DNA), Dick Cotton (mutations), Leo Szilard (nuclear physics), and Maxine Singer (Asilomar). Partly this is a matter of passion, or more precisely how much effort they are prepared to invest in persuading members. The question of political savvy is harder to judge. We have emphasized that politics devolves significant power onto those who set timing and agendas. While this is more art than science, officials – who are often themselves political animals – should have some ability to spot this talent in others.

---

* See Chapter 3.
† This is especially true for academic communities.

Legality. We have emphasized that private power depends on market imperfections. For this reason, organizers will often face significant antitrust issues. The problem, usually, is that they will normally know little if anything about the subject. Despite this ignorance, officials can at least remove roadblocks by insisting on clear answers from their competitions policy colleagues and, where obstacles exist, pressing them to invent imaginative workarounds. More concretely, officials can alleviate concerns that self-governance is a sham designed to facilitate illegal cartels by taking over functions that might raise questions in private hands.*

Legitimacy. Government haſ no business backing community initiatives without some preliminary judgment that they are legitimate. Delaying these determinations is also inefficient, pushing legitimacy debates onto members themselves and draining energy from more practical discussion.

We have argued that modern legitimacy theory is almost always predicated on some loosely conceived "will of the people." We return to this subject in Section 9.5.

9.2. Tactics: Promoting and Influencing Private Initiatives

Every private governance initiative includes two distinct choices: (a) the decision to pursue collective action and (b) agreeing on some specific, substantive solution. Governments that believe in private democracy will often promote the former while remaining agnostic about the latter. This section asks what government can do to encourage private governance without taking sides.

9.2.1. Promoting Collective Action

Kenneth Abbott and Duncan Snidal argue that private transnational governance's "most serious limitation" is an "orchestration deficit" and that shrewd government intervention can dramatically improve the impact, legitimacy, and public-mindedness of private schemes.[30]

Organizing and Convening Meetings. Our coffee and fisheries examples show that private politics often proceeds by developing solutions within

---

* For example, nuclear exporters sometimes share otherwise sensitive customer information by passing it through government intelligence services who would promptly complain if the exchanges went beyond what was needed to detect illicit weapons programs. See Chapter 10.

relatively small groups and then scaling them up through a series of larger gatherings. This implies that political skill mostly lies in picking (and occasionally revising) the timing and sequence of discussions. The fact that private participants almost always see government-hosted meetings as credible and legitimate gives officials significant leeway in deciding when and how to "conven[e] public and private actors,"[31] followed by clever agenda setting and negotiation.[32]

This raises some obvious questions. The first and most insistent is who should be invited? Here, the natural guess is to approach the firms that will ultimately be asked to self-regulate. However, we have argued that private power is almost always enforced by outside anchor firms. This suggests that officials would be better advised to jawbone these downstream firms first. Once they demand a standard, the existence of private politics will follow automatically.

Officials should also recognize that government-sanctioned meetings legitimize participants. This creates a risk that members will attend meetings for their own sake, mouthing pious sentiments while avoiding concrete results. Officials can avoid this, as in our coffee and mutations examples, by promising to walk away from projects that fail to make steady progress. The problem, too often, is that officials might not want to admit failure. One obvious commitment strategy is to specify clear milestones in advance. These should be sufficiently ambitious so that failure is both possible and unambiguous.

Relatedly, officials should be careful to avoid empowering players who join private initiatives for the sole purpose of opposing them. This is admittedly in tension with officials' need to appear evenhanded. One way to manage the problem is to limit meetings to "coalitions of the willing," i.e., those genuinely interested in self-governance. This requires some test to weed out the insincere, most obviously a public promise to abide by a majority (or supermajority) vote when the time comes. A second tactic is to insist that all discussions be open and preferably include journalists. This ensures that would-be dissenters will pay the full cost of any obstruction.

Administrative and Financial Support. We have also seen that governments can provide administrative or financial support.[33] Since officials will normally know very little about the initiative's prospects, these should usually be tied to well-defined milestones. Sponsors in our artificial DNA and coffee examples were particularly effective in convincing members that funding would run out if they failed to make progress.

Our coffee example suggests that hiring or seconding professional staff from government can dramatically increase the chances of agreement.[34]

The overall goal should be to reduce information costs so that members come to meetings with as few unanswered questions or misunderstandings as possible. This communication will be most effective when staff are convincingly neutral about outcomes. Of course, no human can completely suppress his own opinions. That said, there are good reasons for them to try. First, staff members know that they have more influence (and prospects for success) if players trust their neutrality. Second, it is almost always better to settle for a less-than-ideal agreement than to have none at all. Third, neutrality is consistent with long-standing technocratic norms. Finally, successfully achieving a standard will almost always be good for staffers' careers, regardless of detailed content.

Governments and funding agencies may also decide to reimburse organizers and members for travel, meeting costs, legal services, and other expenses. Even small sums can be surprisingly effective: In our artificial-DNA case, companies were invariably generous in donating executives' time but did little or nothing to reimburse out-of-pocket expenses.*

Other Interventions. We close by mentioning some other content-neutral interventions that appear in the literature:

> *Advice.* Officials can point out the potential benefits of self-governance and explain how to organize it.[35] However, this advice will only be influential to the extent that officials possess information that private firms don't have already. We should be skeptical that advice will be worth much in the United States, where officials so far have little personal or institutional experience of self-governance. However, the case is more encouraging in Europe, where diplomats have instigated and managed multiple high-profile initiatives since the 1990s.
>
> *Private/public projects.* Private infrastructure projects sometimes open the door to synergies.[36] The difficulty is that officials usually have no independent way to estimate the value of these opportunities. This

---

* The observation raises a puzzle: why do so many firms happily donate their executives' time while refusing to reimburse even minor expenses? The answer almost certainly depends on the tradition by which most high-technology companies let employees spend ten percent of their time on outside projects. See, e.g., Berners-Lee and Fischetti (2000) at pp. 67–8 (describing Hewlett-Packard). Such systems are inherently honest to the extent that the employee would be working on something else regardless. By comparison, introducing cash payments opens new consumption opportunities like the chance to gather with friends in nice hotels. Firms may also want outside sponsors to contribute support as a signal that they, too, believe that the project has merit and are willing to be associated with it.

suggests that government should require recipients to make matching investments that cannot be recovered unless the project succeeds.[37]

Endorsement. We have emphasized that anchor firms often delegate power to NGOs in exchange for endorsements. This incentive is a fortiori stronger for democratic governments that typically possess significant publicity assets and/or accumulated stores of public trust. In many cases, government's most valuable contribution will be to praise what is praiseworthy.

9.2.2. Taking Sides

Democratic governments often see private politics as just another tool for achieving their own preexisting goals. Interventions typically rely on a mix of coercion and persuasion.

Coercive Methods. Governments have long manipulated private politics through the threat and occasional reality of regulation. This, however, requires some reasonable prospect of intervention. Scholars often discuss the problem as if officials can carry out the threat anytime they want to. But in fact, decisions to develop (or discontinue) regulation consume large amounts of bureaucratic energy so that they only arrive at long intervals. Mid-twentieth-century governments typically followed a default practice of letting industry develop its own standards first and only then deciding whether to adopt, improve, or overrule them.[38] By comparison, maintaining parallel official regulation requires an early decision that private efforts are likely to be insufficient. Even when regulation is not attempted, government can sometimes use praise (criticism) to increase (reduce) the standard's value to participating firms.

Perhaps the key difficulty in these cases is that industry has little or no incentive to volunteer information about which standards are feasible and affordable. The most obvious solution is to plagiarize this information from private standards. Alternatively, government could incite competition among multiple standards on the pattern of our lumber and food examples. Apart from procurement policy, however, government is seldom in a position to do this.

Persuasion. We have argued that private politics proceeds through more or less imperfect transfers of information. It follows that policymakers can influence private bodies by sharing data, particularly where – as in most national security discussions – officials possess more and different information. That said, government officials will need to show patience: while government can influence, industry will almost always have the final word.[39] Traditional mid-twentieth-century practice was for agencies to send representatives who had no power to make binding commitments for either side.[40] Other, less honest methods (for example, deliberately withholding information) will deplete the government's long-run credibility and should be avoided.

9.2.3. Public Responses to Private Regulation: Abstention, Competition, and Codification

Government needs to know whether it plans to ignore, supplement, or displace private regulation. We consider these possibilities in turn.

Abstention. Traditional adversarial models of regulation are expensive: monitoring and enforcement (on the government side) and confirming compliance (on the private side) consume resources that would be better spent on the problem at hand. Furthermore, solutions are almost entirely limited to behaviors that can be specified in advance, even where discretion would be less clumsy and inefficient. Given these disabilities, it will often be true that industry can perform the task more efficiently and at less cost. But this bargain is only available *provided government can trust the private process*. Despite this, government seems to make this calculation surprisingly often, most notably in our food example where the European Union deliberately ceded food regulation to industry.

Regulating in Parallel. Governments have long recognized that private standards can elicit new solutions, demonstrate political feasibility, and generate useful fact discovery. They also offer a testbed for policy experiments that can later be ratified or scaled up,[41] provide first-draft regulation that can more easily be harmonized later,[42] and provide a kind of first offer that regulators can lock in or use to demand upgrades.[43] However, all of these presuppose that private standards exist and are active for significant periods of time. This is unlikely if private organizations expect government to ignore and immediately overrule whatever standards they produce, or if the expected public standard is so close to the private one that the latter's comparative benefits shrink faster than the expected costs of completing it.

Private and public regulations can also interfere with each other. In our artificial-DNA case, for example, the United States used regulatory controls to restrict what American companies can export. But these same restrictions make exports less profitable, sheltering local suppliers that could not otherwise compete with Western competitors' economies of scale. In these circumstances, lower official standards would actually increase the dominance of Western firms and their ability to impose private standards. This suggests that higher regulatory burdens can sometimes lead to *less* security. More precisely, officials should never press private *or* public regulation to the point where Western companies' cost advantage disappears entirely.

Sequential Regulation. In the best case, "state regulators ... and private schemes observe one another, borrow techniques, compete, and otherwise co-evolve over time."[44] This happens naturally for private standards, which must please their constituents or disappear. But government officials can and often do ignore private standards entirely. Simple strategies for fixing this might require agencies to (a) address the adequacy of any private standard that overlaps their missions, (b) explain why additional public regulation is or is not necessary, and (c) act or build from the private standard as necessary.

9.2.4. Enforcing Ground Rules

At some level, government cannot avoid setting minimum standards for private rulemaking or second guessing solutions it disagrees with. We have seen that U.S. courts have long specialized in due-process reviews that emphasize a particular style of information gathering based on noticed hearings and other borrowings from common law. This seldom matters for today's large New Self-Governance initiatives (e.g., coffee, fisheries), which tend to mimic government rules in any case. But many smaller bodies operate as little more than informal committees. Here, basic fairness could include respecting private democratic choices on how many safeguards are required and phenomena like external politics that provide protections not found in conventional government.

Unlike courts, executive branch officials have no formal obligation to embrace private standards. Despite this, they often intervene by promising to adopt private initiatives that respect certain minimal procedural and sometimes substantive parameters.[45] This poses much the same danger as with courts, i.e., that agencies will insist on recreating private governance in government's own image while excluding innovations like external competition which conventional government cannot match and that might sometimes be superior. A less intrusive option might be for agencies to weigh in as trusted intermediaries, for example by announcing when they thought activist charges of greenwashing actually had merit.

Inclusiveness. Traditional federal policy encourages the adoption of private standards from bodies that practice openness and consensus procedures. Modern scholars echo this advice by recommending that government use its influence to make private governance more participatory and increase representation for weak or diffuse groups.⁴⁶ By comparison, our shadow electorate arguments suggest that anchor firms already act democratically while the representation of weak or diffuse groups is set by the credibility of their walkout threats, i.e., their prospects for petitioning presumptively legitimate institutions like the press or traditional government. In either case, intervention could actually make private organizations *less* democratic than before. For example, we saw in Chapter 2 how organizers' efforts to skew organizations against specific players in our artificial-DNA and lumber examples led to external competition and were, to a large extent, self-correcting. In principle, government could have intervened to improve the terms offered to outsiders. But in that case, outsiders would have received more power that the usual walkout threats give them. This result only makes sense if we believe that their outside alternatives (including petitioning the government itself) are unfairly limited.

Transparency. Many scholars have argued that government should use its influence to reinforce private transparency⁴⁷ or even provide information directly to consumers so that they can make their own choices.⁴⁸ This book has similarly stressed that transparency improves private politics by reducing randomness so that outcomes more reliably converge on Downsian results. Strangely, the New Self-Governance has often retreated from this transparency. Since the 1990s, industry standards rules that required open meetings have increasingly been replaced by so-called Chatham House procedures* that report comments without attributing them to specific individuals. The justification is that this encourages communication by people who might not otherwise be willing to voice their opinions publicly. But one might well ask why policymakers would want such a thing. The point of political discovery is not to find out what "opening offers" or "strategic bluffs" players demand in private. Such information is, after all, easy to obtain. Instead, what officials really need to know is the "bottom lines" of what players are willing to accept and/or demand publicly. The Asilomar and artificial-DNA examples suggest that open meetings are a powerful tool for suppressing the kinds of bluff and extreme positions commonly found in private settings.

9.3. Managing External Politics

Some observers have argued that government should intervene in external politics by endorsing or criticizing individual standards, promoting

---

* Sometimes called "not-for-attribution" rules. See, e.g., Berger (2013) at p. 11.

uniformity across schemes, forcing mergers to "ameliorate excessive multiplicity,"[49] or fostering new organizations to compete with standards they dislike.[50] Governments potentially possess various levers for doing this, including verbal endorsements (or condemnations) and public procurement policy.*

Sensible though they are, these suggestions proceed from the assumption that private politics is essentially arbitrary so that outsiders can recognize and overrule instances of unfairness as they see fit. But this position becomes untenable if we accept that private politics is itself a reliable and legitimate way to develop and implement standards that reflect the shadow electorate's preferences. Worse, we have seen that internal and external politics are linked: as our coffee and fisheries examples show, the power that groups ultimately wield in internal politics almost always depends on their ability to organize a competing organization if their demands are not met. This, in turn, depends on their ability to successfully petition courts, agencies, legislatures, and the press. To the extent that we think that the latter bodies are legitimate, power sharing negotiated against the threat of walkout is itself presumptively legitimate. Officials who arbitrarily overrule these results risk making private governance less democratic instead of more so.

It follows that officials should be reluctant to intervene in external politics. One exception is when they can increase transparency, most notably when one or both sides are deliberately dishonest. A second exception, following our discussion in Chapter 4, is in helping members overcome the frictional costs of organizing new standards and disbanding old ones. Probably the clearest case, echoing our forestry example, is where competing standards have largely converged so that further competition seems wasteful. Helping the parties build trust to the point where they can negotiate a suitable armscontrol treaty seems useful in these cases. Even here, however, the parties should normally be left to their own power-sharing formulas.

To the extent that officials do exert pressure, they should be sensitive to the fact that they are substituting their own judgment for that of dissenters who plainly prefer continued external competition despite its accompanying waste.

9.4. Academic Self-Governance

The existing Republic of Science model is rooted in shadow of hierarchy arguments. The challenge is to extend traditional peer review methods for allocating resources to individuals to include issues that affect the collective.

---

* Gulbrandsen (2014) at pp. 76, 86.

9.4.1. Promoting Collective Action

The naive message from Chapters 1 and 5 is that academic self-governance is possible (Asilomar, atomic physics) but easily derailed (mutations, synthetic biology). This suggests that the difference between success and failure will often depend on clever strategy. Here we describe five promising strategies drawn from the historical examples found in Chapters 1 and 5.

Coalitions of the Willing. Probably the most obvious advice is that organizers should attempt collective action only when it is necessary. This is particularly true of community projects that focus on constructing new shared assets. Normally, these start on the Silicon Valley pattern by expanding one member at a time and only later become universal. In this case, dissenters cannot simply block action but must affirmatively persuade others not to join. For their part, organizers should be careful not to give dissenters an unnecessary veto. This means, inter alia, limiting membership to individuals and entities (a) whose work and/or data are actually needed to create the project, (b) who expect to receive net benefits if the project succeeds, and (c) who understand and agree in advance to the outline of a common business model. They should also adopt clear majority or supermajority voting rules so that "constitutional" issues are settled before they become urgent.

Adjusting the Payoffs. We have argued that academic communities' ability to pursue collective action is limited by members' need to preserve trust relations. But the value of trust is ultimately measured by the resources it lets members extract from funding agencies, universities, and private firms. Grant agencies can change this calculus by rewarding members who support self-regulation, most notably by making adherence to private standards a "plus factor" in deciding grant applications.* Agencies can also earmark support for community-wide assets, so that they become still more attractive to members. Resources can also come from nongovernment sources, as in our mutations example, where the shared facility would (a) make members' lives better, and (b) attract commercial support for key players that agreed to drop their opposition and supply paid services instead.

---

* The idea that grant agencies should compensate members injured by regulation was repeatedly suggested in our nuclear physics and Asilomar examples, although nothing seems to have come of it.

Finally, government could imitate external politics by, for example, announcing that they intended to make strong biosecurity standards a "plus factor" in grant awards. This would immediately create strong pressures to create community-wide standards. Government could then decide which of multiple competing standards it ought to reward.

Coordinating Across Agencies. Sometimes it would be enough for funding agencies to get out of the way. This was almost certainly the case in our mutations example, where individual players fought a community-wide facility because they hoped agencies would fund their own individual projects instead. Looking back, it would have been much more sensible for funders to decide at the outset whether collective action was globally desirable and, if it was, to defund competing individual proposals entirely.

Editors' Conspiracies. Editors' conspiracies have provided an important shortcut to academic self-governance since the 1940s. One advantage is that editors largely stand outside the trust networks that link the majority to dissenters. This is because no single scientist can afford to cancel her subscription to *Science* and if she did it would not matter. Furthermore, only aggregate readership counts.* This suggests, as in conventional democracy, that every "vote" is equal. The fact that dissenters usually feel more passionate than the majority counts for nothing.

Moratoria. Finally, the feasibility of self-regulation depends (and should depend) on designing interventions that minimize cost and maximize benefits. One clever way to do this, following several of our examples, is to focus regulation on new research methods before the average member has invested much time and effort in learning them. One countervailing weakness is that our historic examples have almost always promised members clear expiration dates, most notably the defeat of Nazi Germany in our atomic-physics example. This assured members that any sacrifice was only temporary so that they would eventually be able to publish and claim rewards for their work. This pattern poses deep challenges for fields like synthetic biology, where the most natural policy response is to suppress some weapon-related technologies forever.

---

* There are exceptions. The editors of *Physics Today* panicked in the 1960s when recent Nobelist Richard Feynman asked to be removed from their mailing list, hurriedly writing back to ask how they could produce a magazine that "physicists want and need." Feynman responded with a kind of genial irritation that "I'm not 'physicists.' I'm just me. I don't read your magazine ... Maybe it's good. I don't know. Just don't send it to me." Feynman (2005 [1966]) at p. 227.

9.4.2. Making Government a Better Partner

Our examples document recurring instances where officials were either indifferent to private governance or else had no clear idea of how to advance it. The question remains how fast government can overcome this learning curve. The problem is particularly acute in the United States, which has had relatively little experience interacting with private initiatives compared to Europe.*

Promoting Engagement. Our case studies suggest that officials respond to private initiatives with deliberate silence. But if government agencies have formed an opinion on whether a particular private intervention is desirable, it makes no sense to hide this information from citizens. Probably the simplest fix is for agencies to require officials to affirmatively explain why private initiatives that overlap their mission do or do not serve the public interest. A second, longer-term strategy would be to create a political culture that stresses the importance of engaging private initiatives. To some extent this will happen anyway as the number of private initiatives grows. More formal interventions could be achieved through legislation and executive order. This might well be overkill: statements by high-level diplomats and respected bodies like the NAS go a long way toward making the point and may be sufficient. Finally, agencies' hiring and promotion committees should take account of officials' efforts to influence private standards bodies, especially when they achieve results comparable to more traditional methods.

A Transactions Culture. Compared to traditional politics, the New Self-Governance draws much more heavily on how private firms negotiate deals. Officials will have to learn much more about this culture. One place to start is to make government more nimble, providing information and making commitments at something like the speed of business.

Probably the deepest change will be adopting capitalism's backhanded concern for what others think, realizing that community members must consent to new standards and cannot simply be ordered to participate. There will also be various lesser skills, including the ability to draft contracts, a feel for negotiations, and knowing that it is better to let some partners walk

---

* European diplomats were instrumental in encouraging partnership in our coffee, forestry, and artificial-DNA examples. By comparison, American officials kept a careful silence except, arguably, in the Asilomar case. The difference may be due to European governments' comparative lack of resources, which forces them to be more innovative than Washington.

away when the price is too high. Above all, there should be a willingness to try new self-governance initiatives, despite the possibility of failure.

Providing Clarity. Government has an obligation to speak its mind. Our examples contain multiple occasions where officials maintained a careful silence rather than saying what standards they preferred (Asilomar, artificial DNA) or even whether proposed private initiatives were an acceptable use of intellectual-property policy (mutations). The latter case is particularly extreme, since the availability of private funds was intended to provide otherwise unobtainable communal resources – implicitly, at least, to make the NIH's own budget go further. A better approach would be to work out a clear, first-principles framework for deciding which private/public deals (e.g., exclusive contracts, "pass-through" licenses, embargoes on data, or "traffic-builder" sites) would or would not be acceptable. Granting that agencies are unlikely to "bless" specific transactions in advance, a clear explanation of their reasoning and general criteria for deciding when private initiatives are desirable would still be useful.*

Finally, officials should resist the temptation to abdicate controversial policy decisions in favor of peer review.† This is a category error: the Republic of Science is supposed to pass on the quality of science proposals, not economics issues in which scientists have no obvious expertise. Conversely, officials are responsible for making economic and normative choices for their agencies, which will often pay more in the future depending on which policies are adopted. They should not be allowed to delegate such choices even if they want to. Proposals for shared infrastructure face the additional problem that peer review was designed to make incremental decisions on individual proposals. This can tolerate a certain amount of randomness. By comparision, community-wide projects are much more likely to raise fundamental issues about how the field as a whole should develop. Letting randomly selected scientists decide these questions seems irresponsible.

Coordination Across Agencies. In our mutations example, the NIH and other funding bodies repeatedly focused on individual scientists' database proposals. This was a sensible way to compare science merits, but systematically obscured the threshold question of whether community projects/

---

* The U.S. Department of Justice's "Merger Guidelines" also do not set "hard and fast" rules. Nevertheless, corporate America finds them a valuable guide to anticipating what regulators are likely to do.
† I have seen various bodies from Doctors without Borders to the Department of Energy use peer review to decide whether proposed licensing rights in intellectual property and data are acceptable.

private sector support were inherently more desirable than individual projects/public support in the first place. In principle, it would have been better for agencies to consult one another and decide whether a communal approach was desirable, and if so call for a unified approach. Instead, each project was considered in isolation, leading to the worst possible outcome in which no project was ever funded by anyone. Grant administrators should decide early on whether or not to support community-wide solutions. Once they do, they should announce this openly.

9.5. What Scholars Can Do

Public policy, unlike astronomy, is an inherently applied subject. No matter how intellectually satisfying, the goal is to help policymakers solve practical problems. This requires a certain focus: scholars will never match organizers' knowledge of their own individual communities, or officials' knowledge of national security issues. Instead, they should focus on topics where they have some comparative advantage. This generally includes knowing what earlier self-governance ventures have tried. It also includes theory and logical arguments, where scholars – unlike players and officials – have nearly unlimited time to study and sharpen their understanding.

9.5.1. Feasibility and Legitimacy

The deepest problems, as we have seen, concern feasibility and legitimacy. Nevertheless, we can sketch a way forward. The Sherman Act identified otherwise wooly political and philosophical judgments ("power," "legitimacy") with economic conditions that can be objectively measured on the ground. Furthermore, if one self-governance example is legitimate and feasible, then other proposals that are indistinguishable (or systematically superior) from an economic standpoint should likewise be approved. Scholars' research should stress these comparisons and, eventually, emergent judgments about which projects are acceptable.

A Roadmap. Figure 9.1 summarizes this book's arguments by translating questions of feasibility and legitimacy onto objective questions of market structure. To see this, start from the vertex at the top of the figure. Here, perfect competition disables private power so that private firms have no power and no discretion. At least in its original form, the Sherman Act was meant to restore this world. Proceeding clockwise, the second vertex ("Cartel and Monopoly") represents unbridled power by individuals and

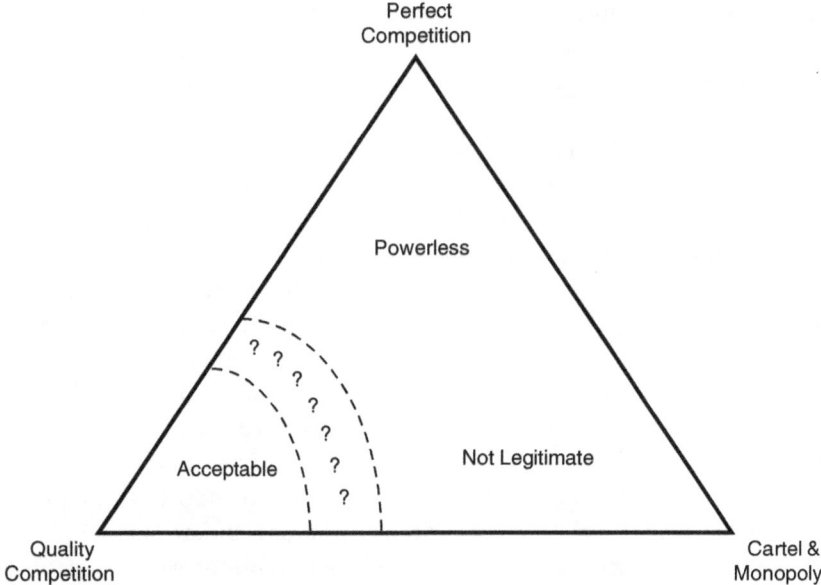

Figure 9.1 Feasibility, Legitimacy, and Market Structure.

cartels. Here, private power is plainly possible but its legitimacy is debatable. One might have thought that putting knowledge in private hands was acceptable so long as the players resembled the broader society or else could be trusted to converge on mainstream results through deliberation. As we recounted in Chapter 7, however, the Sherman Act puts such arguments off-limits in a way that only Congress can revisit. Following the logic of Douglas's *Alcoa* dissent, the very purpose of the Act is to keep power away from tiny, randomly selected elites.

So far we have focused on the extreme right-hand boundary of the figure that spans the continuum from perfect competition to cartels. But this book has stressed a third vertex that Congress did not consider. Here, dominant anchor firms cartelize price competition but continue to compete fiercely on quality. The novelty compared to the two traditional cases is that private firms (a) possess formal power over suppliers but (b) possess little or no choice in how they use it. This leads to shadow electorate arguments in which private governance follows the will of consumers and, implicitly, the people.

Shades of Gray. Of course, policymakers will hardly ever encounter our polar cases in life. Instead, real industries will normally mix all three states so that they fall in the interior of the diagram. This presents the usual line-drawing challenge of deciding when regions are "close enough" to our polar cases to be considered both feasible and legitimate. We label these "Acceptable" in the diagram. In general, the judgment will mix *both* value judgments *and* objective fact.

To be sure, simple market-share arguments are only a starting point for benchmarking legitimacy and feasibility. The dotted lines in Figure 9.1 accommodate this uncertainty. On the legitimacy side, evidence of transparency should normally expand our figure's "Acceptable" region to the right. Probably the most obvious approach would be to ask real consumers what they know about private standards. This, however, could be naive given how little even conventional voters know about politics.[*] A more realistic approach is to assume that shadow electors (like ordinary voters) only pay attention when some controversial incident invites scrutiny. The question then becomes how quickly self-governance organizations respond with new standards.[†] Additional facts will normally include statistical analyses of how quickly firms react to activist criticism and/or standards introduced by rival companies; survey evidence that executives subjectively fear consumer displeasure at levels that are at least roughly comparable to elected officials' sensitivity to poll numbers; the existence of private rules and institutions that make it easy for outsiders to learn about private policies; the existence of widely known and trusted third parties (e.g., the U.S. government, the Sierra Club) that have commented on those policies; the fact that anchor firms have delegated significant power to outsiders; and/or the existence of external politics involving multiple, competing private standards.[‡]

On the feasibility side, policymakers can use the same heuristics, most notably Herfindahl (market concentration) indexes,[§] which facilitate

---

[*] The situation is summed up in an aphorism doubtfully attributed to Churchill: "The best argument against democracy is a five-minute conversation with the average voter." See, e.g., Pollock (2016).

[†] This is similar to the usual political science conjecture that regulation is expensive to monitor, so that even well-organized and -financed bodies like the U.S. Congress prefer to wait for scandals to intervene. McCubbins and Schwartz (1984).

[‡] One ironic possibility is that controversy often increases visibility and consumer awareness. This implies that NGO criticism can sometimes make private organizations *more* legitimate than they were before.

[§] The Herfindahl index is defined as the sum of the squares of the market share (expressed as a fraction) of every firm within a particular industry. Wikipedia (n.d.(d)). The index's principal virtue is that it always increases when market share is transferred from small firms to large ones. This reduces the otherwise bewildering constellation of possible market-share partitionings to a well-behaved if somewhat arbitrary rank ordering. Antitrust experts use this to develop first-glance intuitions about market power within and across particular industries on the understanding that really definitive judgments almost always require additional facts.

first-cut comparisons across industries. Granted that these simple market-share indexes do not fully capture market power, this is a known problem that antitrust judges and juries deal with every day.

Shadow of Hierarchy and Network Markets. So far, we have analyzed New Self-Governance models in isolation. That said, the older shadow of hierarchy and network models described in Chapter 1 continue to exist and can complicate the analysis. Most obviously, government demands for private standards may sometimes help anchor firms impose private governance where it would not normally be feasible. Graphically, this will expand the "Acceptable" region of our diagram upward. This stabilization could reflect government carrots (e.g., endorsements, financial support) as well as sticks (e.g., threats of punishment and regulation). The usual argument, following Chapter 7, is that the resulting regulations will themselves borrow a measure of legitimacy from this intervention. In that case, the "Acceptable" region will expand to the right.

Finally, we have said that many of the oldest and most familiar examples of commercial self-governance exist in "network industries" organized around interoperability standards. These typically possess a strong tendency toward monopoly punctuated by occasional fights to unseat incumbents. Graphically, we expect to find network industries disproportionately drawn to the lower right sector of Figure 9.1.

9.5.2. Bottleneck Issues

Finally, there are important issues where scholars already know enough to contribute. We have seen that self-governance communities spend enormous time debating obscure antitrust and intellectual property questions. On the antitrust side, ordinary members are predictably bewildered. The result is that organizers default to absurd rules like forbidding members to say the word "price" – known to coffee organizers as the "P-word." This plainly makes an already-difficult discussion harder. That would be fine if it solved some real legal problem, but the available evidence is not encouraging. Competent antitrust professionals will normally be able to provide better answers off the cuff, and a really sophisticated analysis in something like an afternoon.

The corresponding bugbear for academic communities is intellectual property. As we saw in our mutations example, U.S. officials are happy to accept commercial money so long as absolutely nothing is given in return. Beyond that, however, they have practically nothing to say about which deals might be acceptable. This is true even though most trades would give academics more resources than they have today. No doubt some of these deals would be unfairly favorable to industry. That said, it is hard to imagine a community-wide deal as burdensome as the current default practice of letting individual scientists sell intellectual property rights for as much as the market will bear. This is just another way in which U.S. grant institutions ignore the special advantages of collective action and shared resources.

10

Extending the Model

Self-governance has a long and impressive history. So far, however, policymakers have seldom done much to encourage it. This is a broad hint that more interventionist policies could dramatically expand the number of initiatives. The question remains whether this would be worth the effort.

This chapter explores the feasibility and effectiveness of extending self-governance to address new social challenges.* Section 10.1 begins by asking whether the basic New Governance model should be expanded to include more industries. We argue that the benefits for most civilian sectors are limited, mainly because so many large anchor firms have implemented private controls already.

The case is radically different for high-technology companies whose products can be used to make nuclear, cyber- and other advanced weapons. Here, most security professionals agree that private self-governance offers one of the best options for strengthening existing safeguards.[1] Sections 10.2 through 10.4 presents a detailed industry-by-industry survey of instances where self-governance is likely to be both feasible and useful.

Section 10.5 addresses the special problems of academic science, with a special focus on innovations that would make research more responsive to the wider society. Finally, Section 10.6 provides a brief conclusion.

10.1. Civilian Technologies

The New Self-Governance exploded onto the scene in the 1990s (tuna, fisheries, lumber) with a slower but still steady expansion (coffee,

---

* Compared to earlier chapters, this discussion is often intensely fact dependent. Readers of a more theoretical bent may prefer to skip these discussions, or else treat them as worked examples of how governance problems should be analyzed in practice.

nanotechnology) since then. The question remains whether still more initiatives are desirable. Here, the threshold question is whether the added difficulty of organizing collective action is worth the trouble. Even where it is, some projects may not be economically feasible or politically legitimate.

Is Collective Action Desirable? Individual anchor firms routinely impose private regulation on their supply chains. Indeed, Conroy (2007)'s encyclopedic account lists dozens of examples.* The question in each of these cases is whether there is anything to be gained by transitioning to collective action, i.e., extending individual supply chain standards to entire industries.

One partial answer is that collective action offers the biggest payoffs where a particular industry controls some specific issue in the way that the lumber industry, say, dominates forestry issues. Absent this overlap, the added expense and uncertainty of organizing collective action suggest that it is normally better to pursue the much simpler task of pressing individual firms to adopt standards.† Beyond that, standards should focus on improving operating methods. As we saw in Chapter 3 (p. 53), efforts to take assets entirely out of production often end in "feel-good equilibria" that rearrange demand with no net improvement.

Feasibility and Legitimacy. Even supposing that collective action is desirable, we still want to know when it is feasible and legitimate, i.e., falls within the "Acceptable" regions of Figure 9.1. While this is ultimately an empirical question, the required market conditions probably exist across much of the U.S. economy. In the words of one leading industrial organization economist:

It is futile to attempt a precise quantitative summary of how much structural monopoly power exists in the whole of the American economy. Suffice it to say that there is a modest amount of activity (not more than 5 percent of GNP) approaching pure monopoly (most of which is subject to government regulation), somewhat

---

* These include, inter alia, private standards for banking and finance, apparel, electronic waste, health insurance, tourism, mining, and medical products.

† To take a somewhat simplistic example, we know that all lumber companies employ accountants. Despite this, organizing collective action across the timber industry would do very little to improve accountants' lives on average. This advice might be different if industry-wide standards were cheap, for example because we expected them to spread sua sponte after one or two big companies adopt them. But while this model *can* work – StarKist's tuna standard is at least arguably an example – it seems to require very concentrated industries. By comparison, our nanotech example shows what happens when big players simply announce standards without making any effort to sign up adherents. Anecdotally, at least, even Web standards seem to require extensive politicking in their early stages. Berners-Lee and Fischetti (2000).

more activity approaching pure competition, and large quantities of oligopoly and oligopolistic competition.*

This makes it reasonable to think that many if not most U.S. industries (a) exercise significant market power but (b) are sensitive to quality competition. It follows that many candidate industry-wide standards would be both feasible and legitimate. The main uncertainty, for now, is that we do not know just how much market concentration is needed. Naively, at least, our examples suggest that StarKist, Unilever, and Home Depot possessed sufficient purchasing power while nanotechnology and coffee did not. Additionally, we have seen that practically all commercial self-governance raises significant competitions policy issues. That said, the fact that authorities never seriously considered intervening in a single one of our examples suggests that any objections are manageable.

Finally, we have emphasized that collective action need not be regulatory and negative. Instead, firms can work together to create shared assets. So long as there are significant returns to scale, these can usually be launched as joint ventures and allowed to accept new members until the entire community participates. Despite significant doctrinal confusion, competitions policy's "essential facility" doctrine generally authorizes this approach.

10.2. Nuclear Technologies: Enriched Uranium

We have already said that many of the most promising opportunities for self-governance involve national security.[2] The next three sections examine the industries and potential interventions that would be needed to limit the proliferation of nuclear, chemical, biological, radiological and cyberweapons. We begin with the classical problem of limiting centrifuge technologies.

10.2.1. Technology and Industry Structure

Nuclear weapons can be constructed using either plutonium or enriched uranium. The uranium path to nuclear weapons requires advanced technologies to purify ^{235}U from its natural abundance of less than 1 percent to 85 percent or higher. Thereafter, constructing unsophisticated bombs is straightforward.[3] In principle, many technologies can be used to accomplish

---

* Scherer and Ross (1990) at pp. 81–2. Twenty years earlier, the book's first edition had pegged the monopoly estimate at between 6 and 7 percent. Scherer (1970) at p. 59. This is in line with the usual perception that U.S. industry has become more competitive since the 1960s. Vogel (2005).

this enrichment, including lasers, robotics, machine tools, and electronics.[4] Here, we focus on the traditional and still most common high-speed centrifuge path.[5]

Centrifuge plants in the legal economy are almost always constructed by large U.S. and European integrator firms that draw on elaborate supply chains for the required skills, trade secrets, and tacit knowledge. But these large turnkey projects are relatively easy for governments to detect and intercept.[6] This has forced modern proliferators to become integrators in their own right.* For example, the network founded by Pakistan's A. Q. Khan signed illicit contracts with a half dozen or so advanced European machine shops while using front companies to purchase more complex hardware from large companies like Leybold.[7] Significantly, proliferation has no clear finish line: given that rogue nations need spare parts to maintain their capabilities, better interdiction could actually *shrink* their capabilities over time.[8]

10.2.2. Detection Options

Company screening programs almost always begin with information generated in the course of normal sales operations. In principle, this enables three distinct strategies for screening sales:

> *Orders.* No matter how clever they are, proliferators must eventually ask for sensitive technologies. This means that many inquiries contain enough information to trigger concerns. Leybold's scandals in the 1980s occurred because employees were either not trained to recognize suspicious orders or, more usually, ignored them to earn bribes or sales bonuses. Most observers agree that subsequent reforms have fixed these problems within Leybold itself. But other firms continue to do very little to monitor orders.[9]
>
> *Customers.* Requests for dual-use technologies normally trigger follow-up inquiries to confirm the customer's claimed identity and intended end use. This generally requires phone calls, checking references, and other labor-intensive investigations. Front companies' ability to resist

---

* In the case of centrifuges, this typically means obtaining roughly one hundred distinct components [Carnegie Fund (2012) at p. 6] from "considerably lower" than one thousand vendors. Wirtz (2010) at pp. 280 and 274. Many of these are only available in the developed world. Carnegie Fund (2012) ("[T]here remains, at least for the foreseeable future, critical components that cannot be obtained outside of key NSG states"). Even South Africa's state-of-the-art industry failed to produce adequate steel rotors. Id. at p. 7.

these probes varies. Some consist of little more than a mailing address or fax machine number. At the other extreme, proliferators sometimes persuade long-established, legitimate firms to place orders on their behalf. In practice, most, though not all, existing front companies are fairly shallow,[10] suggesting that many technology vendors are still easily fooled.

Detecting patterns across orders. Proliferators frequently try to disguise suspicious orders by splitting them into individually innocent subsets.[11] Leybold instituted elaborate in-house databases to detect splitting across its various divisions.[12] Global security expert Gretchen Hund and her coworkers have persuasively argued that industry should extend the same practices across companies, with reasonable precautions to "keep company-specific information confidential for proprietary reasons."[13] One obvious opportunity would be for vendors to trade the names of suspected front companies.[14] Exchanges with government intelligence services, which usually possess very different information, would extend this strategy still further.[15]

As usual, just knowing the possible strategies is not enough. We also need institutions that energetically and imaginatively implement them. The balance of this section asks how well current institutions work and whether private governance can do better.

10.2.3. The Current System

Traditional nonproliferation measures depend on export licenses to overrule otherwise lawful sales. This has at least four weaknesses:

Incomplete coverage. Officials only have enough budget to regulate about 2,400 items.[16] This leaves room for proliferators to purchase lower-quality but still workable substitutes.[17] Most observers argue that regulatory resources would go further if more emphasis was placed on identifying suspicious end users and then reallocating resources to pursue them.[18]

Limited data sharing. Except for Leybold itself, relatively little sharing takes place between the subsidiaries of individual firms,[19] competing firms,[20] or governments.[21] Many companies in Europe and especially the United States still refuse to share data with governments claiming "concerns about loss of proprietary information."[22]

Discarded data. The current system focuses on completed sales agreements. This discards massive amounts of information about inquiries that never get that far, including cases where companies have identified and refused suspicious orders.[23]

Limited expertise. Top-down regulatory models give the last word to government officials, who almost always possess less expertise than the companies they regulate.[24]

Recognizing these weaknesses, reforms since the 1990s have tried to make industry more creative and proactive without fundamentally changing the system's top-down architecture. In practice, this usually takes the form of "catch-all" regulations that require firms to investigate end users' identities and claimed technology needs.[25] However, these depend on the imaginative exploitation of clues, which are almost impossible to specify in advance beyond a handful of very generic "red flags."[26] This places strong practical and especially legal limits on how much creativity occurs in practice[27] and falls well short of an open-ended duty to investigate.[28]

Despite these shortcomings, the current system has made rogue programs significantly more expensive and less capable than their Western counterparts.[29] The question remains whether some other system could do better.

10.2.4. Private Governance Options (A): Regulating Critical Subsystems

Traditional regulation emphasized the danger posed by large integrator companies and, following the Khan scandal, the importance of preventing would-be proliferators from accessing existing supply chains directly. Given that governments had successfully stopped the sale of complete ("turnkey") plants, this necessarily meant interdicting efforts to purchase major subsystems, notably including the elaborate facilities needed to process highly corrosive uranium gas under vacuum.

The Industry. Vacuum technologies are so difficult that only a half-dozen companies worldwide have invested sufficient R&D to develop the hardware, advice, and design services needed to build industrial systems.* During the 1980s, A. Q. Khan successfully targeted Leybold's nuclear design services group to design and deliver vacuum systems for Pakistan's illicit centrifuge

---

* The industry specializes in designing pumps and custom systems for maintaining a high vacuum in harsh chemical environments. The main players worldwide have historically included Leybold, Edwards Vacuum, Atlas Copco, Pfeiffer Vacuum, Ebara Technologies, and Busch Vacuum Pumps and Systems. The first three firms recently consolidated under common ownership.

plant at Kahuta.[30] However, Leybold suffered multiple scandals after its equipment turned up in Iraq following the first Gulf War (1991). At this point, the U.S. Pentagon exerted massive pressure to force top-to-bottom reforms within the company.* As a result, Leybold's screening practices are now far stronger than any competitor's.[31] The question remains whether the rest of the industry can be induced to match or even exceed this effort.

Transitioning to Collective Action. By far the most obvious way to improve security would be to make Leybold's practices universal across the industry. The question is whether the Pentagon and other Western governments have enough buying power to force such rules on the industry's four independent firms. The simplest strategy is just to purchase compliance: given that illicit orders account for 0.1% of all industry sales inquiries, paying even a modest premium for each order should be enough to compensate companies for lost business and fund upgraded screening.[32] The stronger response, following our discussion in Chapter 3, would be to enroll the Pentagon and other government purchasers as "anchor firms" to force negative profits onto any company that refuses to upgrade its practices. For this strategy to work, organizers would have to lock up so much demand that the industry's smallest company could not scrape together enough replacement business to survive.

Beyond their immediate impact, industry-wide standards would also open the door to further improvements. First, government could use procurement policy to reward suppliers that developed new and better screening methods on the pattern of our food, fisheries, and forestry examples. Second, common standards would facilitate subsequent agreements to share information across companies and governments.[33] While business confidentiality objections exist,[34] Leybold has long argued that the most useful transactions data (e.g., quantity, type, date, and origin of inquiries) is not particularly sensitive.[35] The fact that the company routinely shares intelligence with host country intelligence services[36] and the International Atomic Energy Agency (IAEA)[37] provides strong evidence for this assertion.

---

* Leybold's current policies were adopted in response to lost sales, fears of additional backlash, and U.S. Pentagon threats. Albright (2010) at p. 182 (discovery of Iraqi purchases in the first Gulf War produced revenue losses in the United States and Japan); Wirtz (2010) at p. 261 (American government "threatened to blacklist Leybold in the US"); Albright (2010) at p. 182 (United States's threat to blacklist the company); Wirtz at p. 278 (describing allegations that the U.S. Department of Defense told vendors that they "should refrain from doing business with Leybold").

A Grand Bargain? We have argued that the current system could be made more efficient by reducing top-down controls so that firms (a) have more freedom to identify and prioritize investigations of truly suspicious end users[38] and (b) redirect funds currently spent on demonstrating compliance to expand screening efforts.[39] The obstacle, inevitably, is trust. Officials quite reasonably fear that industry would take advantage of relaxed supervision to do even less than they do now.[40]

Despite this, the idea of delegating more authority to private firms remains attractive. By analogy with our Chapter 6 discussion, there are two ways to supply the required trust: (a) increase transparency and (b) build confidence stepwise through multiple repeated transactions. One natural first step would be for companies and government to establish a clearinghouse where mid-level employees on both sides could work together to pool information and identify suspicious patterns. This would immediately let officials judge how enthusiastically companies are pursuing investigations and sharing data. To the extent that these signals are positive, officials could give companies still more freedom to identify warning signs and prioritize investigations.

10.2.5. Private Governance Options (B): Regulating Individual Components

So far, we have focused on using large integrators to police their supply chains. However, it might be better to establish self-regulation within the subindustries themselves. The question, following our Chapter 3 discussion, is whether such rules could suppress dissent. The answer for many industries plainly is "no." The reason, as usual, is fixed costs. Many high-technology components are manufactured by small family businesses organized around a handful of million-dollar machine tools.[41] A single illicit client can typically offer more than enough orders to keep such businesses afloat.[42]

That said, there are significant counterexamples where self-governance might be possible. Probably the most important involves firms which make the exotic materials used to make centrifuge blades, notably including the high-strength carbon fiber compounds which Iran needs to replace its existing centrifuges with more capable Western designs.[43] Here industry structure closely fits our Chapter 3 paradigm in which high, fixed-cost suppliers maintain ongoing relations with a handful of large anchor firms:

> *Suppliers.* Suitable high-performance compounds are currently available from fewer than a dozen firms worldwide,[44] each of which has developed its own proprietary techniques for making the material.[45]

Developing these techniques requires prohibitively large R&D investments that keep new firms from entering the industry. These investments also imply large fixed costs that anchor firms can exploit for leverage on the usual pattern.

Anchor firms. Carbon fiber is used in many different applications, including items like sporting goods. However, about 20 percent of the market consists of aerospace orders,[46] which are overwhelmingly dominated by just two companies, Boeing and Airbus. These companies typically place orders with all major suppliers.[47] At the same time, both companies are large defense contractors and receive government import-export loans. This would presumably encourage them to spend their limited economic leverage on improving nuclear security.

The main uncertainty in this analysis is that about 5 percent of industry capacity is produced in China.[48] At least in theory, China could simply order its manufacturers to supply rogue states even if their overall profits were reduced or eliminated. Whether China would be willing to underwrite such losses is unclear.

10.3. Nuclear Technologies: Plutonium

The second path to nuclear weapons uses a nuclear reactor to transform natural uranium into plutonium. The fuel rods are then chemically separated – a straightforward task – to isolate the bomb material. This makes reactor technologies an obvious chokepoint for nonproliferation policy.* While states can theoretically build their own reactors, commercial technologies remain a valuable workaround for North Korea, Iran, and Iraq.

Industry Structure. The commercial reactor industry produces components and fuel materials worth roughly $40 billion per year.[49] The industry fits our usual New Self-Governance paradigm, with a handful of large anchor firms drawing on an extended network of suppliers:

Anchor firms. The downstream market is dominated by nine large "international technology vendors" or "integrator" firms that build large reactors for making electricity.[50] Two of these firms also sell smaller reactors for scientific research and technology development.[51] The rub is that many integrators are based in China and Russia, whose host

---

* Our survey is not exhaustive. The construction of plutonium weapons is more technically demanding than uranium bombs. A comprehensive discussion would address additional chokepoints, notably including technologies for the milling and precise detonation of explosives.

governments could potentially pressure national firms to dissent even when this led to reduced or negative profits.*

Suppliers. The big integrators are supplied by firms across Europe, Russia, the Ukraine, and China.[52] Most are large manufacturing operations that make forgings, industrial equipment, and the like. These typically feature the kinds of large, fixed-cost investments that generate leverage for anchor firms.

Customers. Unlike our previous New Self-Governance examples, reactors are seldom purchased by individual humans. Instead, customers are almost always governments, government-owned enterprises, or regulated utilities.[53] This suggests a shadow of hierarchy model in which governments exercise power over integrators through purchases, subsidies, and actual and threatened regulation. Large integrators routinely worry about political backlash in the event of an accident or terrorist attack.[54]

Interventions. The nuclear industry has already begun to organize itself. In 2008, the large integrators unanimously agreed to form a Nuclear Power Plant Exporters group to develop industry-wide operating standards.[55],† While discussions have so far been limited to regular exchanges of best practices instead of formal standards,[56] these are said to be "more technically detailed and comprehensive than other, similar industry initiatives."[57] Governments "generally support" the project and were "kept informed throughout the drafting process."[58] The resulting best practices have been publicly endorsed by American, French, and Korean officials.[59]

Challenges. To this point, our reactor industry discussion closely resembles the usual New Self Governance model with governments playing the role of outraged consumers in demanding that anchor firms impose private standards. That said, there are two obvious difficulties with this picture. First, many of the industry's institutional customers are non-Western, with an estimated 57 percent of power-reactor projects planned through 2030 located in China or the former Soviet Union.[60] This strongly suggests that Chinese and Russian integrators do not need Western orders to survive, although they might be less profitable without them.[61]

A second difficulty is that governments have to ensure the physical basis for markets. While the United States was briefly able to cut off fuel for Iran's

---

* Iran has already purchased one power reactor from Russian vendors and has announced plans to buy further units from Chinese sources. World Nuclear Assn. (2016).

† The organization has announced that its membership is open to any firms that become integrators in the future. Perkovich and Radzinsky (2012) at p. 16.

1960s-era research reactor, Iran reputedly used the threat of terrorism – including the rumored assassination of at least one former integrator company executive – to pressure France into honoring previously agreed upon supply contracts.[62] While this book has repeatedly stressed the usefulness of private standards, these are ultimately no better than traditional governments' willingness and ability to protect commercial actors against physical coercion.

The good news, for now, is that the case for stronger standards does not seem to be particularly urgent. Traditional government agencies, notably including the International Atomic Energy Agency, are already said to be "highly effective" at uncovering and deterring illicit reactors.[63]

10.4. Other WMD Technologies

Nuclear weapons apart, all WMD technologies depend on dispersing unusually powerful poisons over large areas.[64] It follows that proliferation threats almost always turn on regulating the manufacturing technologies needed to make exotic chemicals, organisms, or radionuclides in industrial quantities.*

10.4.1. Chemical Weapons

Security experts focus on three basic strategies for making exotic chemicals in bulk. At first blush, at least two appear amenable to self-governance:

Conventional manufacturing. The first and most obvious way to produce exotic chemicals at an industrial scale is to build an illicit plant. In practice, the required design services are available from a large number of small, technology-based firms[65] so that it is possible to start world-class plants anywhere on Earth.[66] Generally low fixed costs make self-regulation an unlikely strategy for controlling these services.

Precursors. Manufacturing challenges are dramatically lower where proliferators can buy precursor compounds from civilian markets. These "specialty chemicals" are almost always made by small to medium-sized private companies in concentrated industries that serve worldwide markets.[67] The fact that members commonly join preferred supplier arrangements[68] within semipermanent supply chains[69]

---

* In principle, there is a trade-off, so that clever new poisons could relax the required manufacturing volumes. In practice, current state-of-the-art compounds required decades of costly animal testing and are unlikely to be improved on. Maurer (2009).

provides a strong hint that anchor firms possess substantial private power.[70]

New manufacturing methods. Chemical manufacturing traditionally required large plants. Over the past decade, however, new "microreactor" technologies have made it increasingly possible to build up capacity from small tabletop units.[71] This R&D-intensive industry has high fixed costs and sells primarily to large chemical manufacturers,[72] suggesting that New Self-Governance methods would almost certainly be feasible.*

10.4.2. Biological Weapons

Pound for pound, bacteria and viruses are far more lethal than chemical poisons. However, naturally occurring organisms are seldom immediately suitable as weapons. This implies that credible WMD programs need access to sophisticated manufacturing and genetic-engineering technologies.

Manufacturing. Large-scale biological weapons attacks typically require hundreds of kilograms of living "agent." This can theoretically be accomplished using competitively supplied, low-technology fermenters. New Self-Governance methods would almost certainly be ineffective in these markets. The case is only slightly better when it comes to more capable computer-controlled fermenters and bioreactors. Given that these technologies can be purchased from literally dozens of small suppliers,[73] the case for governance seems marginal at best.

Genetic Engineering. Research tools that dramatically reduce the time and expense required to create genetically engineered organisms provide a potentially powerful lever for constructing more lethal organisms. Here, artificial DNA is by far the leading candidate. The first and most obvious step would be to revive the private standards which the artificial-DNA industry implemented in 2007–9.† Beyond this, two additional initiatives are worth considering. First, members could agree to **pool customer data**. Current human screening protocols are expensive, sometimes requiring PhD biologists to spend up to two hours on a $10,000 order. At the same time, companies report that they have seen between

---

* Leybold's ownership of industry leader Degussa suggests additional possibilities for leverage. Wirtz (2010) at p. 253.

† The fact that the industry managed to achieve stable self-regulation ten years ago suggests but does not prove that it could do so today. The industry continues to evolve, with a comparatively large fraction of firms entering or exiting each year. . Woodrow Wilson Ctr. (n.d.)

three and five percent of all sequences before.⁷⁴ This makes it reasonable to think many more sequences that are "new" to one firm have already been evaluated elsewhere. Pooling and reusing this data would immediately cut screening costs to that all concerned.* While critics have sometimes claimed that a shared database would compromise customers' business confidentiality,† disclosures would be limited to the fact that an unknown customer had requested artificial DNA at least distantly related to a known GenBank entry. Preliminary discussions with suppliers and select customers in 2007–10 found no evidence that such disclosures would be problematic.⁷⁵ Second, members could also agree to **pool customer data**. The vast majority of synthetic-biology orders are either inherently harmless or come from known or easily verified customers. However, this still leaves perhaps two cases per thousand in which unknown individuals or companies request potentially problematic sequences.⁷⁶ The cost of investigating these orders could be radically reduced by pooling selected customer data in a shared depository.⁷⁷ This would probably include suitably anonymized data summarizing past purchases and/or regulatory interactions.‡

The question remains whether a clearinghouse for sharing customer information would be legitimate and legal. With respect to the former, we note that Big Pharma anchor firms' continued profitability depends sensitively on government regulatory approvals and support for strong patent rights. This suggests a strong case for derived legitimacy in which firms preemptively seek to please regulators.

Finally, a data clearinghouse poses several potential antitrust issues. Probably the most obvious is that any business model that seeks to recoup operating costs through member fees could likewise be used to raise prices. That said, the risk should be manageable so long as fees are modest and/or

---

* Trading judgments would also reassure otherwise suspicious firms that their rivals were not "cutting corners"; promote more coherent policies by requiring members to explain why specific organisms should/should not trigger further investigations; and compile a training set for artificial intelligence programs that could scan the literature to identify still more judgments about which organisms are and are not virulent. The shared database would also be useful for basic science and various technology applications (e.g., pathogen detectors) that government has so far declined to fund. Maurer et al. (2009) at p. 25.
† The literature contains occasional examples of customers asking providers not to compare sequence identities against GenBank (Imperiale 2008) and synthetic-gene companies promising confidentiality as a marketing tool. Minshull (2009).
‡ Companies that have purchased large amounts of DNA in the past, that have previously been investigated and received export licenses, or that have been granted preferred supplier status by one or more anchor firms would normally be considered legitimate.

structured as fixed costs with suitable discounts for small players.* This suggests that antitrust officials would normally approve industry-wide clearing-house projects under the essential facilities doctrine.

10.4.3. Radiological Weapons

Radioactive materials provide our third and final example of WMD based on strong poisons.

Technology and Industry Structure. While there are many isotopes, almost all commercial radionuclides share a common market structure:

> *Suppliers.* Most radionuclides are manufactured in a handful of small research reactors outside the United States.[78]
> *Anchor firms.* Anchor firms include makers of food irradiators, medical and industrial imagers, oil-drilling equipment, and other devices.[79]
> *Customers.* An estimated two million sources are currently located in the hands of end users.[80]

For the most part, the first two levels of the chain are reasonably secure.† By comparison, customers often do a poor job of securing radioactive materials. Probably the biggest gaps concern disposal, especially when the original vendor has gone out of business.[81]

It is natural to ask what anchor firms can do to improve the situation.‡ The novelty, in this case, is that they would have to enforce self-governance *downstream* over consumers. Here the easiest option is to fund education. Experience in the oil and gas industry shows that this is often effective.[82] That said, the strategy works poorly where consumers are numerous and unsophisticated. One obvious alternative is to provide prepaid disposal services directly, including cases where the original manufacturers no longer

---

* Authorities might also worry that an industry-wide threat list would suppress each individual firm's incentives to produce a better list than its competitors. In practice, the advantages of sharing almost always overwhelm this factor.
† Reactor operators generally know their customers, although scholars argue that this could be further improved by sharing more information about suspected incidents and front companies. See, e.g., Ferguson and Lubenau (2003). Anchor firms generally provide good physical security for radionuclides for much the same reason that they guard other high-value materials. Id.
‡ Anchor firms' leverage over suppliers probably varies from market to market. Some isotopes are in perennial short supply, suggesting that suppliers possess significant monopoly power. That said, even monopolists should have no trouble accepting reforms where anchor firms agree to pay for them through higher prices.

exist.* The main practical difficulty is that, contrary to our usual New Self-Governance model, manufacturers would have to police themselves. This would require trust among firms, possibly reinforced by fears of backlash.

Finally, any shared facility would invite antitrust scrutiny. That said, we argued in Chapter 8 that the need to achieve scale economies around a single "essential facility" would provide a strong argument for sharing. The facility could strengthen these arguments still further by structuring itself so that it (a) is jointly owned by all users, (b) is open to all firms that wish to join on nondiscriminatory terms, and (c) structures its participation fees in ways that have a minimal impact on prices and new entry.

10.4.4. Cybersecurity

So far, we have focused on traditional WMD technologies. Since the 1990s, however, many observers have argued that attacks on computer networks ("cyberwar") could inflict similarly large-scale damage. Superficially, one might expect that New Self-Governance-style market interventions would have limited usefulness against solitary hackers or autarkic state teams. On closer examination, however, cyberwar attacks against sophisticated targets (e.g., defense contractors, power grids) require (a) a deep and previously unknown software vulnerability (the "zero-day defect") and (b) a team of skilled hackers and other criminals to exploit it. Successful interventions must target one or both components. We consider each possibility in turn.

Hackers. So far at least, states find it much easier to train infantrymen than to create hackers. This forces them to draw on manpower that was originally trained in commercial settings: for example, Microsoft estimates that about two-fifths (42 percent) of the world's IT professionals work directly or indirectly for its ecosystem.[83] This suggests that Western software companies could use their substantial purchasing power to stop hackers from joining illicit projects in the first place. Probably the most obvious strategy would be to insist that IT workers adopt ethics codes that forbid criminal and/or

---

* A more extreme alternative would be to change industry practices so that manufacturers lease products instead of selling them. This would provide a direct financial incentive for suppliers to keep track of their machines. Even more importantly, customers could simply ask vendors to recover the product when they are done with it. That said, leasing strategies would drastically change market structure with unpredictable impacts on price and output. This raises severe and probably fatal difficulties for competitions policy.

cyberwar activities. Those who refuse or violate the code would be banned from doing further business within the ecosystem. The Chinese Hacker Convention announced such a scheme in 2011.[84]

Even so, such schemes are almost certainly marginal. The reason is that Chinese and Russian markets are only weakly coupled to the West. This suggests that New Self-Governance schemes lack the critical mass needed to impose industry-wide collective action, but would instead follow our forestry and fisheries examples by splitting the market into regulated and unregulated segments. This would presumably leave more than enough talent for illicit programs. For example, China's software companies only sell about one-tenth (12 percent) of their products outside the country,[85] while only about one-third of Russian software developers work for companies that generate significant exports.[86] Meanwhile, the number of IT professionals working for high-end state programs is tiny. This means that Russia can recruit all the specialists it needs from high-end criminal groups.[87] China similarly recruits from its criminal hacker population[88] or purchases contract work from IT professionals in universities and technology companies.[89] These labor pools are almost entirely insulated from Western markets.

Zero-Day Vulnerabilities. Most hacking starts by identifying a vulnerability or "bug" that allows attackers to gain entry to and ultimately manipulate the target computer. In practice, the great majority of these attacks are aimed at poorly-maintained soft targets like home and small business computers. These are almost always based on known vulnerabilities which defenders could have, but failed to, guard against. However, these methods are much less useful against heavily-defended, high-value computers. Here, known vulnerabilities are routinely monitored and "patched." This means that attackers must instead find and exploit previously unknown "zero-day" defects* that the defenders are unaware of and cannot easily defend against. Not surprisingly, zero-days are scarce with fewer than one hundred being discovered worldwide in any given year (Maurer 2017).

As one might expect, criminal groups have organized black markets for buying and selling zero-days. Perhaps more surprisingly, some Western companies have established legitimate markets as well. So far, however, these have been almost entirely limited to law enforcement and intelligence clients that agree to keep vulnerabilities secret. This makes it natural to ask whether it would make more sense for industry groups to organize their own auctions and then reveal their purchases to members or even the

---

* IT specialists use the phrase "zero-day" to connote threats that have never been seen before, i.e. are entirely new on the first day of the attack.

general public. This would drastically reduce the number of zero days and, presumably, attacks. Just as importantly, it would bid up the price of whatever zero days remained. This would immediately make crime less profitable, reducing the number of attackers and, ultimately, the human capital available for cyberwar.*

Crucially, the power of this strategy depends on the industry group's budget for buying defects. This would almost certainly scale with the number of members. In principle, this suggests that collective action might be needed to overcome free-ridership. However, this is not at all obvious. Given that benefits increase with the number of members, some companies could find it in their interest to join regardless. In that case, it might be better for a handful of companies to start the project and then invite other companies to join and expand the effort.

10.5. Academic Initiatives

Finally, we turn to the special governance problems that face non-commercial academic communities. Unlike commercial firms, academic science goes well beyond existing products. This raises the usual Asilomar problem of trying to manage technologies whose properties are both emergent and unpredictable.

10.5.1. Learning and Attention Span

Academic science, more or less by definition, is concerned with cutting-edge technologies that industry has yet to deploy. This introduces two kinds of uncertainty. First, the technology may be so embryonic that no one on Earth really knows what dangers exist. The question then becomes whether so-called experiments of concern could inadvertently reveal unexpected technical possibilities to would-be attackers. Second, emergent technologies often prompt new and unfamiliar normative debates. While most academic self-governance initiatives have deliberately focused on well-defined safety and security† concerns, this ignores the much wider set of issues

---

* Consider a bidding war between our hypothetical industry group and their would-be criminal opponents. Naively, we expect individual defending companies to submit bids up to their expected losses if the zero-day defect was exploited. This suggests that industry groups should include as many members as possible. Meanwhile, criminal groups' ability to bid depends on their expected payoff from mounting attacks. Since the zero-day defect becomes worthless once it is used, criminals will usually bid *more* when there are *fewer* competing groups.

† We exclude meetings convened to discuss issues with no thought of practical action. In 2016, nearly 150 synthetic biologists attended a closed meeting at Harvard Medical School

(e.g., cloning humans) that trouble scientists and the public more generally. The difficulty in these cases is not so much that the public opposes research, but that what little opinion does exist is lightly held.

The defining feature of both problems is that neither the goals nor technical options are well defined. Here the usual town hall methods in which community members meet once to vote on solutions are almost always inadequate. Instead, institutions must have sufficient attention span to accumulate learning and evolve solutions over time.* Over the past five decades, scientists have experimented with several strategies for doing this:

> *Meetings and honey pots.* The simplest institutions provide a forum for unstructured conversations between scientists on the Pugwash pattern. This transparency is already useful: most illicit weapons programs have historically been discovered through informal scientist-to-scientist gossip or confessions.† These could potentially be enhanced by collective announcements that members have an ethical obligation to investigate and report dangerous behaviors.[90] Finally, collective action projects in which members work together to build and use shared assets reward scientists for working together. The resulting repeat interactions would increase transparency still further.
>
> *Advice portals.* Scientists have a long tradition of seeking formal and informal peer review before performing controversial experiments.[91] Recent initiatives have formalized this model to include online advice "portals" where expert panelists (usually including both scientists and security experts) write formal opinions on the advisability of specific experiments.[92]

---

to discuss the ethics of synthesizing an entire human genome, including such questions as whether the technology would open the door to creating designer humans or replicating celebrities. See, e.g., Pollack (2016). Unlike biosafety or biosecurity, the desirability of these goals is unclear and cannot be resolved through evidence-based argument.

* Anglo-American "common law" courts provide the most familiar example of how such systems are supposed to work. At every step, judges are asked to say how a specific set of facts should be resolved, fully disclose their reasons, and explain how those reasons are consistent with or require the modification of previously announced rules. For the classic argument in favor of common law courts, see Holmes (1882).

† Science gossip has historically played a leading role in unmasking covert weapons programs. See, e.g., Richelson (2006) (use of academic gossip to track Nazi atom bomb project); Harris and Paxman (describing how covert Russian and South African biological weapons programs were unmasked by scientist-to-scientist contacts). The method is particularly effective since illicit WMD programs typically employ dozens if not hundreds of scientists. This increases the odds that at least one defector will eventually step forward to expose the program.

Clearinghouses. Science self-governance currently operates mostly in response to controversial experiments and the threat of bad publicity. In principle, at least, it would be better to create groups that periodically scan the literature to identify and develop solutions for emerging issues.[93] To the best of my knowledge, no group has yet attempted this.

The main limitation of these projects, at least for now, is attention span: each new case tends to be decided without regard to previous decisions, so that general principles never emerge. Two partial solutions would be to (a) provide staff to help scientist panels write more explicitly reasoned opinions that address earlier decisions and (b) publish suitably anonymized opinions so that outside scholars can criticize wrongly decided cases and extract coherent principles from those that remain.

10.5.2. Editors' Conspiracies

We have seen that journal editors have acknowledged an obligation to withhold dangerous papers from publication. Following our atomic physics example, the next logical step would be to replace individual conscience with collective institutions for deciding which papers should and should not be published.* This, however, would require some method of enforcing committee decisions against editors who choose to publish anyway. In principle, at least, it may be enough for community members to speak out when editors fail to cooperate.

10.5.3. Building Infrastructure

So far, we have concentrated on the negative aspects of academic self-government, i.e., community efforts to limit or suppress experiments that would otherwise take place. However, self-government also opens opportunities to build new collective assets that promise to make most members better off. The postwar grants system, with its reflexive focus on individual laboratories, has systematically ignored such projects. The result is that projects that require participation by large numbers of workers are typically conducted commercially or not at all.[94]

Our mutations example shows how private initiatives can begin to fill this gap. Database projects are especially promising candidates for shared

---

* The fact that there are fewer editors than scientists to organize makes this option particularly attractive.

assets because they are relatively cheap to build and operate. This means that they can be organized by scientists with little, if any, outside support. More ambitious projects could then attract commercial funds by improving academic data in ways that provide direct financial benefits to biotech.

The main obstacle, so far, is that business has almost always limited itself to quasi-charitable "consortia" that pay academics to pursue work they already want to do. Much more private money would be available if scientific communities were prepared to offer new features that boost commercial research. One obvious opportunity is that commercial firms usually find it prohibitively expensive to collect, rationalize, and unify academic biology data. This suggests that industry should be willing to pay academic biologists to create better data sets in the first place.

10.6. Conclusion: The Limits of Private Power

This chapter has focused on specific cases. We close by asking whether the various hypothetical initiatives discussed above suggest more general intuitions.

Probably the first and most basic lesson is simply that New Self-Governance models are worth considering. Granted that many industries – notably hackers who deal in cyber threats and companies that manufacture nuclear components – fail to meet the New Self-Governance paradigm laid down in Chapter 4, others (e.g., chemical weapons and carbon-fiber technology) look promising. Second, organizers should remember that collective action is not always necessary. Particularly for shared-asset projects, it is often enough to start with joint ventures and let increasing returns to scale attract more members over time. Finally, market interventions are strongly attenuated for industries that produce so-called durable goods (e.g., radionuclides) that remain useful for years and even decades. That said, many assets (e.g., centrifuges, reactors) require a steady stream of replacement parts to remain functional.

Other lessons are specific to national security. First, New Self-Governance strategies cannot deny technology to states that are able to construct their own military capabilities without resorting to markets. Even so, details matter: for example, China and Russia often use markets to find new zero-day vulnerabilities. Second, national security problems seldom involve industries close to consumers, although large chemical and pharmaceutical companies present a partial exception. This suggests that, legitimacy will normally turn on traditional shadow of hierarchy or backlash arguments more than shadow electors. Finally, private governance, like markets themselves, requires physical security. For example, we should not expect private actors to resist the kinds of physical threats that Iran has sometimes employed in the nuclear context. That said, such extreme levels of intimidation would challenge conventional governments as well.

Conclusion

I said at the start of this book that policymakers can have much more self-governance if they make the effort. The main barrier was conceptual – learning some history and replacing lazy shibboleths and half truths with careful, evidence-based models. Readers who have come this far can judge for themselves whether I am right.

But if you agree with me, then the real work can begin. For scholars, that mostly means building better theories. It is very unlikely that I have said everything worth saying about the economic origins of private power or the private politics that sits atop it. On the empirical side, we need to know much more about when the New Self-Governance is feasible. This will require detailed evidence of how often anchor firms are able to enforce "preferred" price and quality terms on their supply chains.

Finally, and perhaps most intriguingly, we know almost nothing about the extent to which markets force anchor firms try to anticipate and honor public opinion. Here several research strategies look promising. On the one hand surveys and case studies can explore the extent to which individual executives fear and feel constrained by mass opinion. On the other, industrial organization studies can catalog the industries where the economics favor self-governance. This will normally include markets where prices have been convincingly cartelized but quality competition continues to keep retailer profits low.

The next step will be for government to put this knowledge to work. Here the deepest changes are cultural: Officials will need to learn and internalize the logic of deal-making and private standards. One basic step is to be honest about what constitutes success. Simply convening discussions and "stakeholders' meetings" is not enough. Instead, each meeting should make measurable and articulable progress toward concrete standards.

In the long run, self-governance will come no matter what we do. The Cold War era, when the U.S. government could do business entirely by regulation and treaty, is long gone. Meanwhile, the concept of private self-governance is so insistent that – sooner or later – officials will learn to use it. But that is not good enough. Americans should judge their leaders by how quickly and shrewdly they embrace it.

Notes

Preface

1 Maurer and Scotchmer (1999).
2 Berners-Lee and Fischetti (2000), p. 75 et passim.
3 Lok (2009) (quoting FBI agent Ed You).
4 For some typical examples, see, e.g., Luongo and Williams (2007) at p. 459 (urging governments to tap "the power of market-based mechanisms to address ... proliferation challenges"); Tucker (2010) (self-regulation of biology technologies would be "a better approach"); Noble (2013) (praising self-regulation by artificial gene manufacturers); Carnegie Fund (2012) at p. 10 (calling for more self-governance by nuclear reactor manufacturers).
5 Luongo and Williams (2007) at p. 470.

Introduction

1 Locke (2013) at p. 22.
2 Id. at p. 20.
3 Gulbrandsen (2014) at p. 75 (private regulatory programs do not derive rulemaking authority from governments); Vogel (2010) at p. 473 ("A defining feature of civil regulation is that its legitimacy, governance and implementation are not rooted in public authority"); Bernstein and Cashore (2007) (private bodies "derive their authority from interested audiences, including especially those they seek to regulate, without asking governments to force adherence to their rules"); Roberts (2012) at p. A-209 (private communities "govern their interests without the direct involvement of the state or its subsidiaries); Abbott and Snidal (1996) at p. 346 ("'[G]overnance' need not mean 'government.' Many international standards emerge and operate through private market interactions. Others are created within formal organizations, but are still orchestrated by private actors).
4 Private power is, of course, based on government to the extent of "rules governing contract law to ... property rights ... play an important background role." Cashore (2004) at p. 510. However, this does not imply that government is more fundamental

than markets, since it is equally true that traditional government would be impossible without the private sector. See, e.g., Stringham (2016) at pp. 193–205.
5 Bernstein and Cashore (2007) (private standards are not "voluntary," but instead "designed to create binding and enforceable rules"); Auld et al. (2010) at pp. 4–5 (private governance has an ambition to enforce "prescriptive 'hard law' specifying "particular on-the-ground behaviors").
6 Bernstein and Cashore (2007) at p. 349.
7 Webb (2004) at p. 382. The available evidence suggests "… preliminary if not conclusive evidence of real change." Examples include environmental programs, protection for apparel workers in developing countries, sustainable forestry, privacy programs, voluntary safety standards, and voluntary online merchant reliability programs. Id.
8 Bernstein and Cashore (2007); Vandenbergh (2006/2007) (governance "generally relates to group decision-making to address shared problems," or purposes and functions usually associated with government).
9 Wirtz (2010) at p. 277.
10 Weber (1919).

1 Prelude: Self-Governance to 1980

1 Putnam (1896) at pp. 199–201.
2 Vogel (2010) at p. 474 (distinguishing between older technical standards that "lower the transactions costs of market transactions" and New Governance initiatives designed to impose "civil regulations [that] require firms to ameliorate negative externalities transactions").
3 See Furger (1997) (describing how Underwriters Laboratories and Factory Mutual regulated product design and manufacturing operations).
4 Lytton (2014) at pp. 543–4.
5 Id. at p. 557.
6 Id. at pp. 546–7.
7 Id. at p. 555.
8 Id. at p. 544.
9 Id. at p. 567.
10 Id. at p. 559.
11 See King and Lenox (2000) (describing private regulation of books and movies); Garvin (1983) (describing private standards for doctors, lawyers, and other professionals).
12 Doyle (1997).
13 Gupta and Lad (1983); King and Lenox (2000) at p. 716 (pollution and public safety standards); Furger (1997); Büthe (2010(c)) at pp. 15–6 (broadcasters and Canadian medical devices). Firms sometimes hope to preempt mandatory regulation and more stringent standards. Maxwell, Lyon, and Hackett (2000) Firms sometimes adopted voluntary standards in hopes of preempting more stringent mandatory regulations. Maxwell, Lyon, and Hackett (2000); Prakash (2000).
14 King and Lenox (2000); Sinclair (1997) (private greenhouse-gas regulation in the United States and New Zealand); Fiorino (2010) at p. 19 (EPA-sponsored initiatives); Pizer, Morgenstern, and Shih (2008).

15 Gupta and Lad (1983) (quoting AHAM's technical director: "[T]he only way to avoid government regulation is to move faster than the government [through] judicious self-regulation"); see also, Lenox and Nash (2003) at p. 344 (quoting Kodak's CEO: "[I]f industry doesn't take the lead on this issue, government will"); Büthe (2010(c)) at p. 20 (nuclear power self-regulation was motivated by desire to preempt government standards).
16 Shiell and Chapman (2000).
17 *Fashion Originators' Guild, Inc. v. FTC*, 312 U.S. 457 (1941).
18 Id. at p. 462.
19 Id. at p. 461. FOGA's standards attempted to outlaw copying of original dress designs, contributing funds to retailers' ad campaigns, offering certain discounts, holding special sales, selling directly to the public, selling in private homes, and holding fashion shows where no merchandise was purchased or delivered. Id. at p. 463.
20 Id. at pp. 461–2.
21 Id. at pp. 461–2.
22 Designs were protected in two private registration offices. *Fashion Originators' Guild, Inc.*, 312 U.S. at pp. 462–3.
23 FOGA hired professional shoppers to find stores carrying pirated designs. Id. It also conducted periodic audits and assessed fines against members who sold to blacklisted retailers. Id. at p. 463.
24 Id. at p. 462–3.
25 Id. at p. 465–6.
26 Dixon (1978) at p. 2; Farrell and Saloner (1988) at p. 235.
27 Hamiliton (1983) at pp. 463–4.
28 Id. at p. 464.
29 Dixon (1978) at p. 18.
30 Hamilton (1983) at p. 463.
31 Id. at pp. 463–4.
32 Id.
33 Id.
34 Russell (2014).
35 Dixon (1978) at p. 19.
36 Recurring themes include a somewhat vaguely defined concept of non-literal "unanimity" that can be satisfied even when some members dissent (Dixon (1978) at p. 19) and the principle that committees should be "balanced" so that no single economic interest dominates (Hamilton (1983) at p. 462).
37 Abbott and Snidal (2009).
38 Id.; See also, Abbott, Green, and Keohane (2016) ("Internally, organizations like FSC are developing new forms of constituency representation and decisionmaking").
39 Economists who model "shadow of hierarchy" dynamics argue that private preferences enter outcomes as a (generally modest) correction to official policy. See, e.g., Dawson and Segerson (2008).
40 See, e.g., Varian and Shapiro (1999); Katz and Shapiro (1994); Scotchmer (2004).
41 For the standard account of "network markets" see Varian and Shapiro (1999).
42 Büthe (2010(a))
43 Berners-Lee and Fischetti (2000) at p. 110.

44 Mayer and Gereffi (2010) at p. 14.
45 Scarpa (1999). For a less formal but more historically informed account, see Lytton (2014) at p. 541. Lytton and McAllister (2014) explore the same issues from an empirical standpoint.
46 This section follows the definitive account in Weart (1976).
47 Id. at p. 23.
48 Id. at p. 24. Fermi was easily the most prominent nuclear physicist in the United States.
49 Id. at p. 24.
50 Id. at p. 24.
51 Id. at pp. 25–6.
52 Id. at p. 26.
53 Id. at p. 28.
54 Id. at p. 28.
55 Id. at pp. 28–9.
56 Id. at pp. 29–0.
57 Id. at p. 30.
58 Heilbron and Seidel (1990) (describing how universities and medical foundations paid for early Big Physics facilities).

2 Legacy: The New Self-Governance

1 The usage follows Roberts (2012).
2 Gereffi (1994, 1999).
3 For an encyclopedic account, see Conroy (2007).
4 Mayer and Gereffi (2010) at p. 6. ("By the mid-2000s ... industry-wide codes of conduct proliferated and became more transparent.")
5 Meidinger (2008); Gereffi (1994, 1999).
6 Golan, Kuchler, and Mitchell (2001) at p. 155.
7 Id.; Carman (2012). For a detailed description of "Dolphin Safe" standards, see Earth Island Institute (n.d. (b)).
8 Golan et al. (2001) at pp. 155, 158. See also Conroy (2007); Gulbrandsen (2009); Carman (2012) (Big Three market share is "more than 80%"); European Cetacean Bycatch Campaign (n.d.) (same).
9 Golan et al. (2001) at p. 155; see also Earth Island Institute (n.d.(a)).
10 Earth Island Institute (n.d.(a)); Conroy (2007) at p. 46; European Cetacean ByCatch Campaign (n.d.).
11 European Cetacean Bycatch Campaign (n.d.); see also, Earth Island Institute (2011) (Earth Island performed 493 inspections in 2011).
12 Conroy (2007) at p. 46 (2007); European Cetacean Bycatch Campaign (n.d.) (dolphin mortality in the Eastern Tropical Pacific fell from "80–100,000 annually in the late 1980's, to under 3,000 dolphins annually today").
13 Prokopovych (2012).
14 Golan et al. (2001) at p. 158.
15 Id.; 16 USC § 1385 (2006). Earth Island even called for a consumer boycott of the U.S. standard. Earth Island Institute (n.d.(c)).
16 Golan et al. (2001) at p. 158; Carman (2012).

17 GFSI (2016) (GFSI was organized in response to "several high-profile recalls, quarantines, and the associated negative publicity").
18 Wellik (2012); cf. Fuchs and Kalfagianni (2010) at p. 13 (retail supermarkets are highly concentrated, with the top-five retailers controlling at least 70 percent of the market in most countries); Havinga (2008) (supermarkets' purchasing power makes "retail food safety standards obligatory for all who want to stay in the [UK] market"); van der Kloet and Havinga (2008) (the "purchasing power of supermarkets makes retail food safety standards in fact obligatory for many manufacturers"); Verbruggen (2013) at p. 521 ("Given that food retail markets are strongly concentrated today, particularly in Europe, the refusal of a retailer to contract with a non-certified supplier can signify the end of the supplier's business").
19 See Wellik (2012) (describing example in which adoption of common standard reduced required audits from seventeen to two).
20 Fuchs, Kalfagianni, and Arendtsen (2009) at p. 41 (uniform standards "induce supplier participation"). Small supermarkets that adopted an idiosyncratic standard might not attract any bids at all. Id.
21 van der Kloet and Havinga (2008).
22 Havinga (2006) (all but one UK supermarket uses the British Retail Consortium Standard).
23 Havinga (2006) (Dutch retailers adopted the British Retail Consortium Standard because suppliers were "more prepared to accept that").
24 Fuchs, Kalfagianni, and Havinga (2008) at p. 356.
25 Id.
26 Fuchs and Kalfagianni (2010) at p. 4 (EurepGAP was formed to harmonize conflicting national standards).
27 van der Kloet and Havinga (2008). As of 2012, GLOBALG.A.P. covered 100,000 producers in ninety-eight countries. Lebel (2012) at p. A-128.
28 Fuchs et al. (2008) at p. 356 (some estimates run as high as 99 percent.). Id.
29 van der Kloet and Havinga (2008).
30 Fuchs et al. (2009) at p. 36; Havinga (2008) at p. 524; GFSI (2014) ("[T]he GFSI Guidance Document provides a template against which food safety management schemes can be benchmarked and recognised as science-based, contemporary, and rigorous. It is a tool which fulfills one of the main objectives of GFSI, that of determining equivalency between food safety management systems"). The initial founders included 400 manufacturers, retailers, and service providers. Members included Walmart, Coca-Cola, McDonalds, Delhaize, and Sodexo. Lytton and McAllister (2014) at p. 320.
31 Supermarket giants Tesco, ICA, Metro, Migros, Ahold, Walmart, and Delhaize adopted the same four standards two years later. GFSI (n.d.); see also Wellik (2012) (certification consultants report that "WalMart's backing was huge for GFSI").
32 Fuchs et al. (2009) at p. 36.
33 Fuchs et al. (2008) at p. 362 (SQF); Havinga (2008) (IFS and BRC).
34 Fuchs et al. (2008) at p. 361 (GLOBALG.A.P. membership was expanded in 2001 to include roughly equal numbers of suppliers and retailers); GLOBALG.A.P. (n.d.) (GFSI's current board is evenly divided between suppliers and retailers); but see Kalfagianni (2010) (arguing that GFSI is dominated by retailers and large suppliers). By comparison, the big regional standards are still dominated by retailers. Fuchs et al. (2008) at p. 362 (SQF); Havinga (2008) (IFS and BRC).

35 Campbell and Le Heron (2007) at p. 142.
36 Djama, Fouilleux, and Vagneron (2011) at p. 196.
37 Id. at p. 196.
38 Fuchs and Kalfagianni (2010) at p. 20 (GLOBALG.A.P. has prevented national food-safety regulation from developing in Europe).
39 Id. at p. 11; Commission Regulation 178/2002, 2002 O.J. (L 031) 1 (EC). I am grateful to Karl Riesenhuber for explaining EC regulations in this area.
40 Fuchs and Kalfagianni (2020) at p. 18, n. 76 ("We cannot be certain that ... governments would have implemented superior regulation ... in comparison to retail standards"); Büthe (2010) at p. 5 (even the United States finds it hard to regulate certain kinds of food).
41 Havinga (2008) at p. 515 (describing Dutch regulatory pressure on private food inspections). Some scholars speculate that the U.S. Food and Drug Administration avoids promulgating official standards so that it can blame producers if an outbreak occurs. Lytton and McAllister (2014) at p. 292 and n. 12.
42 Lytton and McAllister at pp. 323–4. Imported foods account for just 15 percent of the American market. Id. at p. 327.
43 Büthe (2010) at pp. 13 ("[M]onopsony gives the retailers tremendous power," which they have used to regulate "large parts of the food supply of the Western world"), 1 (regulators "exercise power in a Dahlian sense"); Fuchs and Kalfagianni (2010) at p. 10 ("[R]etail food corporations" have "the power to govern"); Havinga (2008) at p. 529 (private regulation is "detailed" with "a high degree of intervention curtailing freedom of regulated firms").
44 The new standards drove many small farmers out of the market. Fuchs et al. (2009) at p. 30; Fuchs and Kalfagianni (2010) at p. 8; Campbell and Le Heron (2007) at p. 142. GLOBALG.A.P. currently certifies 130,000 farms in 120 countries. GLOBALG.A.P. (nd(b)).
45 Lytton and McAllister at pp. 290–1. For a comprehensive review of studies measuring the impact of food standards, see Lebel (2012). Two studies report that GLOBALG.A.P. has improved hygiene, health, and safety on farms for at least some workers. Id. at p. A-136; but see Havinga (2006) at pp. 528–9 (European retailers still prioritize price over safety while private enforcement is lax).
46 Hale (2011) at p. 309.
47 Howes (2008) at pp. 81, 83.
48 Wortman (2002).
49 Bernstein and Cashore (2007) at p. 353.
50 Unilever Corp. (2003) at p. 8.
51 May et al. (2008) at p. 14; Hale (2011) at p. 311; but see Howes (2008) at p. 88 (quoting certification fees ranging from $20,000 for small community fisheries to "perhaps $200,000 to $300,000 for the largest fisheries," not including the cost of new or revised practices).
52 Conroy (2007) at p. 216; California Environmental Associates (2012) at p. A-72.
53 Gulbrandsen (2009) at p. 658 (MSC certification has worked poorly in fisheries that lack quotas or operate in unregulated waters). The corresponding figure for non-whitefish was just 6 percent. Id.
54 Id.
55 Unilever Corp. (2003) at p. 10. Other fisheries tried to block this strategy by certifying established species. Auld et al. (2007) (Alaskan and Canadian salmon).

56 California Environmental Associates (2012) at p. A-75.
57 Unilever Corp. (2003) at p. 4.
58 Unilever Corp. (2003) at p. 6.
59 Unilever Corp. (2003) at p. 4 (Unilever promised to "work closely" with suppliers to change fish-stock management and fisheries practices); Gilmore (2008) at p. 270 (describing Unilever discussions with Alaska pollock fishery: "At the time, Unilever was one of the largest, if not the largest, whitefish buyer in the world" and bought a large fraction of its needs from the Alaskan market).
60 Unilever Corp. (2003) at p. 4. For similar policies by other retailers, see Howes (2008) at pp. 81, 95 (Walmart), 95 (Marks & Spencer), and Cummins (2004) at pp. 92–3 (Sainsbury's).
61 Unilever Corp. (2003) at pp. 6, 8.
62 Howes (2008) at p. 95; Steering Committee (2012) at p. 18 (only about 10 percent of MSC-compliant fish are labeled as such).
63 van der Kloet and Havinga (2008) at p. 15 (listing supermarkets, fish restaurants, and seafood retailers that "prefer to sell MSC labeled seafood"); Conroy (2007) at p. 219 (Walmart "has notified all its seafood suppliers that it is seeking to purchase only MSC-certified wild-caught fish."); Cummins (2004) at p. 93 (discussing Sainsbury's commitment to "work with their suppliers" to move toward sustainable fish); Howes (2008) at pp. 95 (describing Walmart's promise to work with suppliers, WWF, and Conservation International to develop sustainable practices) and 96 (many UK retailers have announced a "'buyer preference' for MSC-certified fish").
64 Hale (2011) at p. 312.
65 Howes (2008) at p. 88.
66 May et al. (2008) at p. 33; but see Cummins (2004) at pp. 89–90.
67 The Principles let certifiers design 80–100 unique "performance indicators" for each fishery. Cummins (2004) at p. 87; cf. May et al. (2008) at p. 20 ("[T]here is a huge diversity of strong opinions around as to what is an acceptable consequence of fishing. Not all of these opinions are free from self-interest").
68 Fisheries typically interview several certification companies before selecting one. The discussion often includes asking what reforms will likely be needed, the chances of successful certification, and expected fees. Howes (2008) at p. 88. The fact that fisheries make the biggest upgrades after pre-assessment but before certification shows that certifiers usually honor these promises. Id.
69 Cummins (2004); Gilmore (2008) at p. 277. Gilmore claims that certifiers deliberately manipulate scores so that they can order "additional, time-consuming steps to placate" NGOs. Id.
70 May et al. (2008) at p. 24 ("Every certification to date has been subject to comments typified by 'We support certification but not for this fishery' and 'If this fishery is not certified then the MSC is doomed'").
71 Gulbrandsen (2009) at p. 655; see also United Nations Food and Agriculture Organisation (2009) at pp. 131–50.
72 May et al. (2008) at p. 24 (describing U.S. federal fishery managers' anger at the "overly critical assessment report" and how MSC "sought an audience to calm the waters").
73 Gulbrandsen (2014) at p. 81.
74 Id. at p. 84.
75 Id. at pp. 85–6.

76 Id. at p. 81.
77 Id. at pp. 84–5.
78 California Environmental Associates (2012) (surveying efficacy studies); Gilmore (2008) at p. 277 (Alaska's pollock industry joined MSC because it was already one of the world's best fisheries); Bush et al. (2012) at p. 289 (WWF refused to license its panda logo to fishermen who claimed to follow higher standards than MSC); Gilmore (2008) at p. 272 (Alaska pollock fisheries pursued MSC certification as a hedge against NGO and U.S. government pressure); Ward (2008) at pp. 217–18 (Australian fishermen cited MSC approval as evidence that further reforms were unnecessary).
79 Hale (2011) at pp. 312–13; Howes (2008) at p. 94 (describing MSC reforms in South Georgia Chilean sea bass fishery); see also Steering Committee (2012) at pp. 61–4 (describing reductions in seabird mortality and bycatch).
80 Gulbrandsen (2009) at p. 658.
81 Hale (2011) at p. 313; see also, California Environmental Associates, "A Literature and State-of-Knowledge Review of Fisheries Certification and Standards," pp. A-70–A-87 in Steering Committee of the State of Knowledge Assessment of Standards and Certification, *Toward Sustainability: The Roles and Limitations of Certification* (2012) at p. A-77 (certification is limited to North America, Europe, Oceana, and Japan and accounts for roughly 40 percent of all wild seafood).
82 International Coffee Organization (2013) at p. 3.
83 Id. at p. 13.
84 Künkel et al. (2008). Large coffee companies also worried that volatile prices would encourage rich-nation consumers to try new beverages. Id.
85 4C Association (2010) at p. 1; Künkel et al. (2008).
86 The main examples were Fairtrade, Organic, Rainforest Alliance, and Utz Certified. Auld et al. (2010) at pp. 13–14; Kolk (2013) at p. 325.
87 4C Association (2007) at p. 5; Conroy (2007) at p. 2.
88 Federal Ministry for Economic Cooperation and Development (Deutsche Gesellschaft für Technische Zusammenarbeit).
89 Much of the actual work was done by the German government's Bundesministerium für Wirtschaftliche Zusammenarbeit und Entwicklung (BMZ) as implementation body. 4C Association (2007) at p. 4.
90 Kolk (2005) at p. 229.
91 See Künkel et al. (2008).
92 4C Association (2010) at p. 10 (producer demand for higher prices); Kolk (2006) (big corporations agreed on the need for high, stable green coffee prices); Künkel et al. (2008) (government wanted 4C to replace Cold War–era price agreements).
93 Kolk (2012) at pp. 82, 84; Beisheim and Kaan (2010) at pp. 131–2 (noting Western coffee retailers' fear of potential scandals).
94 Künkel; see also 4C Association at p. 5 (4C's roaster members commit to buying increasing amounts of coffee from verified producers over time and reporting their purchases annually).
95 Künkel et al. (2008); Auld at p. 221. 4C continues to work closely with the ICO and its board includes an ICO representative. Auld at p. 229.
96 See Künkel et al. (2008); see also Beisheim and Kaan (2010).
97 Kolk (2005) at p. 229.
98 Kolk (2012) at p. 81.

99 Steering Committee (2012) at p. 68; Conroy (2007) at p. 2 (describing how Starbucks used preferred-buyer agreements to impose private standards on its suppliers).
100 Künkel et al. (2008) (Noting that many earlier standards had "been developed by experts with little or no input from those who are intended to comply with the standards").
101 Id.
102 Id.
103 Id.
104 Id.
105 Id.
106 Id. The new partners included the European Coffee Federation, the British Development Corporation, and the Swiss State Secretariat for Foreign Affairs. Id. Additional sums were donated by the Flanders International Cooperation Agency. 4C Association (2007) at p. 22.
107 Id. (describing negotiations that led to 4C's code and governance documents).
108 4C Association (2007) at pp. 2–4; Auld et al. (2010) at pp. 13–14; Künkel et al. (2008).
109 4C Association (2007).
110 Id. at pp. 2, 6.
111 Künkel et al. (2008).
112 4C Association (2007) at p. 8.
113 Künkel et al. (2008).
114 4C Association (2007) at p. 2.
115 Sara Lee left 4C in 2008 but remained a de facto ally through its association with the 4C-compatible Utz Certified standard.
116 Kolk (2012) at p. 84 and Table 2.
117 4C Association (2010) at pp. 4, 12; Auld (2010) at p. 220.
118 4C Association (2010) at p. 2.
119 Id. at p. 6.
120 Id.
121 Id. at p. 7 ("The Strategic Plan 2015 defines clear activities and measures to better balance supply and demand in the medium-term").
122 Kolk (2012) at pp. 329–30 (supply stood at 14 percent while demand was just 6 percent, of which 4C supplied an estimated 2 percent.).
123 4C Association (2010) at pp. 1, 2.
124 Humphreys (1996) at p. 1.
125 Auld, Cashore, and Newsom (2002) at p. 11; Bernstein and Cashore (2004) at p. 47; Meidinger (2006) at p. 50.
126 Meidinger (2006) at p. 51. See also, Auld, Cashore, and Newsom at p. 272; Conroy (2007) at p. 244; ITS Global (2011).
127 Gulbrandsen (2014) at p. 79; Vogel (2010) at p. 484 (Austria abandoned its official labeling scheme after developing countries complained to the WTO); Büthe (2010(c)) at p. 22 (the Austrian government supplied "a substantial part" of FSC's start-up costs after developing countries and domestic business interests forced it to repeal its tropical hardwood ban); ITS Global (2011) at p. 12 (documenting Austrian, Dutch, and Mexican financial support); Purbawiyatna and Simula (2008) at p. 44 (documenting Danish, German, Dutch, and Swedish financial support).

128 See, e.g., Auld et al. (2002) at p. 272 (FSC's governance structure was designed to "eliminat[e] business dominance" over standards); Bernstein and Cashore (2007) at p. 39 (FSC voting is structured so that business occupies a permanent minority).
129 Forest Stewardship Council (2012) at p. 7; Meidinger (2006) at p. 79 (describing FSC's "expansive vision" of a "just forest").
130 Sasser (2002) at p. 240.
131 Home Depot Corp. (n.d.(a)).
132 Home Depot. (n.d.(b)).
133 Meidinger (2006) at pp. 50, 57–8; Home Depot (n.d.(a, b)) (promising "preferential treatment" to certified products and companies that practice responsible forestry).
134 Home Depot (n.d.(a, b)); Meidinger (2006) at p. 58.
135 Auld et al. (2002) at p. 273.
136 Vandenbergh (2007) at p. 920 (2005 figures). This dominance was even greater in the 1990s when Home Depot and Lowe's accounted for 71 percent of the market. Sasser (2002) at p. 237.
137 Meidinger (2006) at p. 58 (early adopters included Lowe's (U.S.), Homebase, B&Q, and Sainsbury's (UK), and Kingfisher (Europe and Asia)); Bernstein and Cashore (2004) at p. 51 (FSC was supported by "B&Q, Sainsbury, Centex, and German publishing companies.").
138 Harris and Germaine (2002) at pp. 264–5.
139 Two of the top-five homebuilding firms adopted sustainability standards in the early 2000s. Sasser (2002) at p. 237.
140 See Bernstein and Cashore (2004) at p. 51 (German publishing companies); Sasser (2002) at p. 237 (select U.S. and Netherlands publishers); Purbawiyatna and Simula (2008) at p. 80 (Kingfisher, IKEA, and Walmart).
141 Purbawiyatna and Simula (2008) at p. 83 (stocking different certification standards "is costly and sometimes physically impossible due to a lack of space"); see also Mayer and Gereffi (2010) at pp. 9–10; Conroy (2007) at p. 244; Gulbrandsen (2014) at p. 82 (arguing that government purchasing preferences are amplified by "suppliers' preference for relatively simple supply chains.").
142 Sasser (2002) at p. 241.
143 Purbawiyatna and Simula (2008) at p. 82 (quoting Walmart: "[S]hifting to sustainable timber has not added one single penny to the price of our tissue"); Meidinger (2006) at pp. 76–7 ("very few" retail consumers are willing pay price premiums); ITS Global (2011) at p. 8 (same); Vogel (2010); Cashore and Auld (2012) at p. A-94 (studies show that consumers' willingness to pay certification premiums is unstable and limited to select niche and high-quality products sold in North American and European markets); but see id. (describing early claims that U.S. consumers paid premiums of 3–5 percent for certified lumber).
144 Purbawiyatna and Simula (2008) at p. 11 (while retailers and consumers give "preferences to sustainably produced timber," tropical suppliers do not believe that they are willing to pay premiums high enough to cover costs); Cashore and Auld (2012) at p. A-94 ("Little is known, however, about where these certification costs accrue within a market supply chain"); Auld, Cashore, and Newsom (2002) at p. 277 (only 30 percent of surveyed landowners expected to receive a premium for certified wood).

145 Harris and Germaine (2002) at pp. 259 (most forest-products firms purchase their wood fiber needs from small brokers and loggers on the open market), 263 (quoting self-reported data: roughly one in five small mills belongs to SFI); International Wood Markets Group (2013) (top-five producers produce 31 percent of U.S. lumber; top-twenty produce 58 percent).
146 Overdevest and Zeitlin (2014) at p. 35; Cashore and Auld (2012) at p. A-103 ("it is probably accurate to say that certification has led to a separation of markets (certified versus noncertified) but has not reduced the demand for noncertified products – even for the worst producers).
147 See, e.g., Meidinger (2006) at p. 77 (reporting consensus among researchers that forest certification has produced "improved environmental and social practices in many parts of the world"); Mechel (2006) at p. 10 (2006) (analysis of Corrective Action Requests suggests that FSC has improved forest management); Mayer and Gereffi (2010) at p. 2 (FSC has improved specific forestry practices); Steering Committee (2012) at pp. 62–3. (Corrective Action Requests are numerous and require substantive, on-the-ground changes about half the time; case studies in Sweden, the United States, Norway, Bolivia, and Brazil all document at least limited improvements).
148 Steering Committee (2012) at pp. 69–70.
149 Id. at p. 71.
150 Brown and Zhang (2005); see also Purbawiyatna and Simula (2008) at p. 31 and n. 11 (forest certification costs $1/hectare every five years and $.50/hectare during off years, while chain of custody costs $500 per assessment regardless of acreage); Auld et al. (2002) at p. 278 (more than 80 percent of forest owners believe certification expenses, including administrative burdens and operating restrictions, are significant).
151 Bowen (2012).
152 Cashore and Auld (2012) at p. A-93 (compliance costs mainly consist of auditing fees, audit preparations, and changes to operations.).
153 Meidinger (2007) at pp. 54, 67–8 (SFI staff drafted initial standard in consultation with member companies and focus groups).
154 Berg and Cantrell (1999).
155 Published figures vary. Compare Berg and Cantrell (1999) at p. 33 ("15 forest products companies were asked to leave the association ...") with Brown and Zhang (2005) at p. 2056 ("as many as 17" forest-products companies let their memberships lapse) and Conroy (2007) (ten members withdrew and "another 15 were dismissed.").
156 Brown and Zhang (2005) at p. 2056.
157 Gulbrandsen (2014) at p. 82.
158 ITS Global (2011) at pp. 5, 11.
159 Purbawiyatna and Simula (2008) at p. 45; see also Steering Committee (2012) at pp. 62–3 (developing nations Zambia, Brazil, and Malaysia also resisted private certification).
160 Id. at p. 11.
161 Id.; PEFC (n.d.(b)) (PEFC "Is tailored to the specific needs of family- and community-owned forests" that lack the scale required to finance and achieve FSC certification).

162 SFI (2009) at p. 1.
163 Mechel et al. (2006). Many retailers purchased wood interchangeably from both standards. Bernstein and Cashore (2004) at p. 40.
164 ITS Global (2011) at pp. 12, 20; FSC (n.d.); see also Meidinger (2006) at pp. 53–4.
165 ITS Global (2011) at 12, 20; Meidinger (2006) at 53–54 (business originally possessed fewer votes and board members than other chambers).
166 Id. at 69. FSC's voting rules codify the consensus principle by guaranteeing that none of the three stakeholder groups can be outvoted. This is qualified by other rules that specify "no chamber shall show sustained opposition." Mechel, et al. (2006) at p. 43; cf. Visseren-Hamakers and Glasbergen (2006) at pp. 411–12.
167 FSC (2012) at p. 3 (reporting adoption of a revised, organization-wide principles standards based on a 75 percent majority).
168 See, e.g., ITS Global (2011) at pp. 12, 20; Meidinger (2006) at p. 54.
169 ITS Global (2011) at p. 20.
170 ForestEthics (2010) at p. 4 (moderate ecologist resigns from SFI board); Meidinger (2006) at p. 57 (Nature Conservancy resigns from SFI board); Meidinger (2007) at p. 69 (industry members unanimously walk out on FSC working group tasked with developing local standards, causing the parent organization "consternation" and a new "understanding" that industry acquiescence "is expected").
171 Meidinger (2006) at pp. 73, 83.
172 Auld, Bernstein, and Cashore (2008) at p. 427.
173 Meidinger (2006) at p. 55 (describing how AF&PA responded to critics by making SFI independent with a more inclusive board of directors); Brown and Zhang (2005) at p. 2056 (describing how SFI created an MSC-style expert advisory board to recommend improvements); Meidinger (2006) at pp. 67, 82 (describing how SFI added notice-and-comment requirements to rule making); Berg and Cantrell (1999) at pp. 33–4 (describing how SFI introduced mandatory audits for members who display its logo on packaging); United Nations Economic Commission for European Forest Products, *Annual Market Review* 2011-2012 (2012)(hereinafter "UNECE") at p. 115. (describing how SFI implemented FSC-style chain-of custody procedures so that certified wood can be tracked from forest to store); Mechel et al. (2006) at pp. 10–12 (FSC introduced detailed "chain of custody" procedures and required public disclosure of audit results).
174 Mechel et al. (2006) at pp. 48–9 (describing how individual PEFC national bodies authorized old-growth logging and the use of genetically modified organisms); Rotherham (2011) at p. 16 (describing PEFC's Sustainable Forest Management Requirements).
175 Meidinger (2007) at p. 62; Wortman (2002); Visseren-Hamakers and Glasbergen (2006) at p. 412.
176 Mechel et al. (2006) at p. 49; Purbawiyatna and Simula at p. 36.
177 Cashore (2002) at pp. 517–18.
178 FSC International at p. 10. FSC has also debated lower worker safety standards for Canadian, Russian, and tropical suppliers. Meidinger (2007) pp. 66–7.
179 Cashore (2002) at pp. 517–18. For additional detailed descriptions of convergence, see Overdevest and Zeitlin (2014) at p. 34.
180 Meidinger (2006) at p. 78.
181 Cashore and Auld (2012) at p. A-93.

182 Mechel et al. (2006) at pp. 11–12 (adding that activists claim that PEFC's regulations are looser in practice).
183 Overdevest and Zeitlin (2014) at p. 35. Meidinger (2007) at pp. 56, 78; see also PEFC (2004) (FSC and PEFC group compete but also influence each other); Overdevest and Zeitlin (2014) at p. 33 (arguing that FSC/PEFC competition has led to "substantial adjustment on both sides"); Auld et al. (2008) at p. 427 (citing FSC/PEFC rivalry as an example where "[e]volution may occur through competition with other third-party certification systems").
184 Cashore and Auld (2012) at pp. A-92, A-93. For a detailed comparison, see Stephen M. Maurer, "Public Problems, Private Answers: Reforming Industry Self-Governance Law for the 21st Century," *DePaul Business and Commercial Law Journal* 12(3): 297–360 (2014)].
185 Meidinger (2008).
186 See, e.g., Rotherham (2011) at p. 610 ("As only 9% of the world's forests have been certified it seems that the focus should be on expanding the number of FSC-approved standards and increasing the area of certified forests and not on attacking the competition").
187 Cashore and Auld (2012) at p. A-102.
188 UNECE (2012) at p. 113.
189 Purbawiyatna and Simula (2008) at pp. 22, 76, 78 (citing statements by Switzerland, the European Parliament, and the G8).
190 Meidinger at p. 59.
191 UNECE (2012) at p. 113.
192 Cashore and Auld (2012) at p. A-95.
193 Purbawiyatna and Simula (2008) at pp. 75 (Denmark), 75–6 (Belgium), 76 (Switzerland), 76–77 (Germany accepts FSC and PEFC but not SFI) and 76 (Japan and New Zealand); Mechel, Meyer-Ohlendorf, Sprang, and Tarasofsky (2006) at p. 6 (UK); PEFC (n.d.(a)) ("PEFC certification is a standard of choice for public timber procurement [in] the United Kingdom, Germany, and Japan"). Many governments let suppliers show sustainability without reference to any standard at all. Purbawiyatna and Simula (2008) at p. 10 (UK, Denmark, Germany, Japan, France, and New Zealand).
194 Mechel, Meyer-Ohlendorf, Sprang, and Tarasofsky (2006) at pp. 7–8 (Constraints include the WTO Agreement on Government Procurement and various EU directives).
195 Purbawiyatna and Simula (2008) at p. 75.
196 Gulbrandsen (2014) at p. 82; Cashore and Auld (2012) at p. A-95 ("As far back as 2001, a survey found that public bodies accounted for 14% of the global demand for certified wood").
197 Meidinger (2006) at pp. 78, 82–3.
198 Meidinger (2006) at p. 78. (certification generates a "considerable amount" of information that "influences government programmes, perhaps most often through small, incremental ... changes in enforcement practices or interpretations of regulations, but sometimes also through changes in formal requirements").
199 Cashore and Auld (2012) at p. A-97.
200 Cashore and Auld (2012) at pp. A-98–A-99.
201 UNECE (2012) at p. 109; Rotherham (2011) at p. 610.
202 UNECE (2012) at p. 109.

203 Id.; Cashore and Auld (2012) at pp. A-87 (most U.S. and European industrial forests are certified), A-96 (France, Germany, and UK account for nearly half of all certificates).
204 Cashore and Auld (2012) at p. A-96 (Russia has certified 29.7 million hectares; Canada has certified 40.2 million).
205 Cashore and Auld (2012) at p. A-97; Purbawiyatna and Simula (2008) at pp. 17, 21, 27 (small and medium-sized tropical suppliers cannot afford certification); Mechel, Meyer-Ohlendorf, Sprang, and Tarasofsky (2006) at pp. 14, 37 ("comparatively huge investments" will be needed to meet FSC standards in the tropics). PEFC faces the additional obstacle that national forestry associations are rare in the tropics. Purbawiyatna and Simula (2008) at p. 28. At current growth rates, it will be eighty years before half of the world's forests are certified. UNECE at p. 109.
206 BASF (n.d.).
207 Id.
208 DuPont Chemical and Environmental Defense Fund (2007).
209 Id.
210 Id.
211 Fiorino (2010).
212 Insight Investment (n.d.); Fiorino (2010).
213 Responsible NanoCode; Friedrichs (2007); Oxonica Corp. (2007); Bowman and Hodge (2009). Other founding members include BASF, Unilever, Johnson & Johnson, Smith & Nephew, Oxonica, Tesco, and various consumer groups and NGOs. See Responsible NanoCode (n.d.).
214 Responsible Nano Code (2008).
215 Id.
216 Friedrichs (2007).
217 Responsible NanoCode (2008).
218 Id.
219 Id.
220 Chilvers and Macnaghten (2011) at p. 26 (the NanoCode "has been dormant for a number of years").
221 Nanowerk (n.d.).
222 Bowman and Hodge (2009).
223 IG DHS (2008).
224 Nanowerk (2008).
225 Maurer, Fischer, Schwer, Stähler, and Stähler (2009).
226 Grushkin (2010). Since most synthetic biology companies are privately held, market share information is uncertain.
227 Tucker (2010); Check Hayden (2009b) at p. 22; Garfinkel, Endy, Epstein, and Friedman (2007).
228 Maurer (2011(a)).
229 Maurer and Rutherford (2007).
230 Tucker (2010); see also Engelhardt and Maurer (2013) (quoting former U.S. Ambassador Robert Mikulak: a synthetic-biology treaty would take "decades").
231 Maurer, Fischer, Schwer, Stähler, and Stähler (2009) at pp. 7–8.
232 Cello, Paul, and Wimmer (2002) (polio); Tumpey et al. (2008) (1981 influenza). I have heard academic gossip that Wimmer consulted colleagues before proceeding.

233 Aldhous (2005) (reporting that two of three gene-synthesis companies contacted failed to screen orders).
234 Compare Grushkin (2010) (synthetic-gene industry earned about $50 million in annual revenues) with Milne and Tait (2009) at p. 108 (individual "blockbuster" drug revenues routinely top $1 billion per year).
235 Tucker (2010) (Blue Heron reports that it screened customers before 9/11 but only began screening sequences afterward. By "the mid-2000s" "many suppliers of synthetic DNA in the US and Europe" had begun screening orders.)
236 For a detailed comparison of the practices at five unnamed companies, see Bernauer et al. (2008) at pp. 20–2. Violators that fail to screen are widely known in the industry. Maurer et al. (2009); Maurer and Fischer (2010) at pp. 41–7.
237 Maurer, Fischer, Schwer, Stähler, and Stähler (2009) at pp. 19–20.
238 May (2010) (quoting IDT lawyer Damon Terrill); Eisenstein (2010) at p. 1225 (noting that self-regulation would "provide reassurance to large corporate clients that represent [artificial-DNA manufacturers'] bread and butter").
239 Hayden (2009(a)).
240 Maurer et al. (2009) at p. 2.
241 Maurer et al. (2009).
242 Fischer and Maurer (2010).
243 Eisenstein (2010) at 1126 (while it is currently impossible to accurately predict gene function based on sequence, better automated methods "should nevertheless be within reach" and "could greatly improve" screening); Committee on Scientific Milestones (2010) (automated virulence judgments are a "distant goal").
244 Fischer and Maurer (2010).
245 Maurer, Fischer, Schwer, Stähler, and Stähler (2009) at p. 14.
246 Id. at p. 2.
247 Id. at p. 6. Most gene-synthesis companies earn one-third of their revenue from just three customers and two-thirds from their top-ten customers. Id. at p. 18.
248 AstraZeneca (2008) at p. 4.
249 Tucker (2010).
250 Blue Heron, Geneart, Codon Devices, Coda Genomics, BaseClear, Bioneer, and IDT. Tucker (2010). As previously noted, Geneart was a German company that operated substantial U.S. facilities.
251 Tucker (2010). ICPS later worked with the FBI to create procedures for reporting suspicious orders. Id.
252 ICPS (2006) (promising that members would "work together to develop technologies that improve safety and security in synthetic biology").
253 ICPS (2006) (promising to "work with governmental organizations to help facilitate the creation of a governance framework and associated safety protocols to foster an appropriate regulatory environment for the synthetic biology industry"); see also Bügl et al. (2007).
254 Bügl, Danner, Molinari, Mulligan, Park, and Reichert (2007) at p. 627 (arguing that the ultimate solution would involve using "validated software tools to check synthesis orders against a set of select agents or sequences to help ensure regulatory compliance and flag synthesis orders for further review").
255 Id. at 628.
256 Id. at 628–29.

257 Id. at 629 (conceding that harmonized screening standards and advanced software are "unresolved issues")
258 See, e.g., Maurer, Lucas, and Terrell (2006).
259 Bügl et al. at pp. 627 (caption to Figure 1), 629.
260 Id.
261 International Association Synthetic Biology. Members included ATG:biosynthetics, Biomax Informatics, Entelechon, Febit Holding, and Sloning BioTechnology.
262 Bernauer et al. (2008). Workshop participants included representatives from Eurofins MWG, Sloning, ATG:biosynthetics, Febit, Entelechon, TESSY, Information Services to Life Science, Geneart, Craic Computing, and Integrated DNA Technologies. The first seven companies were IASB members. Id.
263 Craic wrote the industry's dominant screening software and then partnered with ICPS on possible next-generation software.
264 These included the author, IT specialist Jason Christopher, and Terry Taylor.
265 Bernauer et al. (2008)
266 Maurer et al. (2009)
267 Tucker (2010).
268 Id.
269 Anon. (2008) (praising IASB's Code).
270 Biological Weapons Convention States Parties Conference (December 2009).
271 Fischer and Maurer (2010).
272 Tucker (2010). See also DNA2.0 presentation slides, FBI Building Bridges Conference, San Francisco, Cal. (Aug. 4–5, 2009) (on file with the author). Geneart and DNA2.0 presented the same slides at an NAS workshop several weeks later.
273 Maurer and Fischer (2010) at pp. 41-7.
274 Id.
275 See, e.g., May (2010); see also Check Hayden (2009b) at p. 22 (quoting IASB executive Markus Fischer: the list-based proposal is "a kind of lowest-common-denominator idea," that is "frankly a little bit naïve and dangerous"); Tucker (2010) (Geneart/DNA2.0's proposal was "less capable" than existing screening methods); HHS (2009) at p. 9 (acknowledging threat from non-Select Agents but arguing that a "comprehensive list ... is not currently feasible").
276 Grushkin (2010) (DNA2.0, Geneart, Blue Heron, GenScript, and Integrated DNA Technologies "... disappeared into a glass-walled bar on the roof of the Mark Hopkins Intercontinental Hotel" and "made an agreement" on the spot).
277 Check Hayden (2009b).
278 Grushkin (2010).
279 Tucker (2010) (DNA2.0 and Geneart held "a series of secret meetings with other large gene-synthesis providers"); Grushkin (2010) at p. 44 (DNA2.0 executive described the discussions as a "secret pact").
280 Grushkin (2010).
281 Tucker (2010); Check Hayden (2009a) (quoting DNA2.0 CEO Jeremy Minshull: "I think what the IASB is doing is great, but we do have a perspective about the scale of the gene-synthesis industry, which helps us to decide what are practically implementable solutions").
282 Grushkin (2010). For the next two years, members took pains to avoid mentioning IASB or its standard either publicly or in print. The author saw this as late as a January 2010 AAAS meeting.

283 Check Hayden (2009a).
284 Lok (2009).
285 IASB (2009, 2014).
286 IASB (2014) (quoting company executives).
287 Lok (2009) (Blue Heron and DNA2.0 claim that their practices "don't differ much from what's outlined in the IASB code").
288 Lok (2009); Check Hayden (2009a).
289 IGSC (2009).
290 Id. at p. 4.
291 U.S. Department of Health and Human Services (2009).
292 Id.
293 Tucker (2010) at p. 23; Fischer and Maurer (2010); Dando (2010).
294 Tucker (2010).
295 HHS (2010) at p. 1.
296 Id. (remarking that "many providers have already instituted measures to address these concerns" and that "ongoing development of best practices in this area is commendable and encouraged"); Eisenstein (2010) at p. 1225 (final guidance left room for companies to apply "equivalent or superior" screening standards); Tucker (2010).
297 Check Hayden (2009b).
298 Wadman (2009).

3 Commercial Self-Governance I: Private Power

1 Auld et al. (2010).
2 Bernstein (2011); Cafaggi and Renda (2012) at p. 13.
3 Maitland (1985) at p. 135.
4 The phrase "anchor firms" is due to Roberts (2012).
5 Bernstein and Cashore (2007) at p. 349; Auld, Bernstein, and Cashore (2008) at p. 423 (power is organized through the market's supply chain); Gunningham and Rees (1997) at p. 392 ("[S]uccess is most likely where there are: a small number of firms in each sector; domination of each sector by large firms; sectoral associations able to negotiate on behalf of their members; and a sympathetic business culture"); Steering Committee (2012b) at A-184 (demand from large retailers is often "the driving force for supplier firms to certify").
6 Locke (2013); Mayer and Gereffi (2010) at p. 4 (self-governance is most favored where "lead firms enjoy some measure of market power over suppliers and some ability, therefore, to affect their behavior."); Steering Committee (2012) at p. ES-14 ("Certification systems have expanded most rapidly when market-leading firms choose them as a means to incorporate improved practices into a supply chain").
7 Mayer and Gereffi (2010) at p. 15 ("In an effort to reduce transaction costs and spread risk, lead firms are promoting rationalization of their global supply chains, with an emphasis on a smaller number of larger, more capable suppliers in a handful of strategically selected countries").
8 Mayer and Gereffi (2010) at pp. 14 ("[P]rivate governance should be greatest when there is a powerful lead firm in a stable value chain"); Steering Committee (2012) at pp. 52 and A-189. ("Large corporate purchasers ... can use their power to pressure

suppliers and other smaller, dependent players in their supply chain to certify. The more streamlined a supply chain, the more influence a buyer can wield").
9 This approximately corresponds to Gereffi, Humphrey, and Sturgeon's (2005) typology of "captive" and "relational" supply chains.
10 Mayer and Gereffi (2010) at p. 8.
11 Id.
12 Bernstein and Cashore (2007); Mayer and Gereffi (2010) at p. 5. ("The genius of this approach was in recognizing that the industrial governance structures established by lead firms to manage their global supply chains could also be leveraged to achieve social and environmental objectives").
13 Vogel (2010).
14 Mayer and Gereffi (2010) at p. 11 (describing apparel standards: "[O]nce a critical mass of lead firms found it in their individual interests to adopt private codes, those firms had a collective interest in convergence on common standards."); Steering Committee (2012) at p. 54 ("[S]hifting a whole commodity sector onto a certified footing involves collective action by major buyers to put pressure on intermediary traders and suppliers").
15 Mayer and Gereffi (2010) at p. 11 (common standards "… minimize the compliance costs of suppliers who sold into more than one chain" and also allow "greater monitoring efficiency").
16 Steering Committee (2012) at p. A-183 (industry-wide standards promote "shared learning and legitimation of new practices").
17 Id. at p. A-185.
18 Id. at p. A-184.
19 Id. at p. A-183 (arguing that benefits scale nonlinearly for "safety in numbers" standards that consumers recognize).
20 Id. at p. A-185.
21 Maitland (1985) at p. 136.
22 Roberts (2012).
23 Mayer and Gereffi (2010) at pp. 9, 10, 14.
24 Mayer and Gereffi (2010) at pp. (i), 13–14.
25 Steering Committee (2012) at p. A-189.
26 Abbott and Snidal (2009) at p. 556. The authors add that the phenomenon remains "uneven," particularly for global schemes which try to include "Southern voices … [but remain] dominated by Northern elites." Id.
27 Vogel (2005).
28 Steering Committee (2012) at p. ES-6.
29 Id. at p. 71.
30 Id. at pp. 71 and ES-6.
31 Id. at p. ES-14. The committee adds that standards do sometimes persuade firms "to make significant changes in their practices …" Id.
32 See, for example, AstraZeneca's ethics code in our artificial-DNA example.
33 For a formal microeconomic model of how private standards empower voluntary contributions by consumers, see Besley and Ghatak (2007).
34 Vogel (2005).
35 Besley and Ghatak (2007); Kotchen and van 't Veld (2009).
36 Vogel (2010) at p. 479. (Firms that market to consumers "are particularly risk averse," although some suppliers also "value public approval and dislike negative media attention.")

37 Fischer and Lyon (2014) (surveying literature on market choice where standards are few in number or poorly trusted and understood by consumers).
38 Maitland (1985) at p. 136.
39 Id.
40 Maurer and von Engelhardt (2013) at p. 20.
41 Id. at p. 16–17, 21–22.
42 Abbott and Snidal (2009) at p. 526 ("For example, collaboration between NGOs that favor high labor standards and firms that are willing to accept higher standards but prefer self-regulation may result in a joint standard that is more effectively implemented than a pure NGO scheme (because of the firms' business expertise and management capacity) and more legitimate than a pure industry code (because of the NGOs' normative expertise, commitment, and independence)").
43 Maurer and von Engelhardt (2013).
44 Jackson presents several examples from the literature. Marseilles fish buyers who purchase more than two tons per month do more than 80 percent of their business with one seller at a time. Jackson (2008) at pp. 330–1. Similarly, 60 percent of New York clothing firms send at least 50 percent of their business to just one manufacturer. Id. at pp. 329–30.
45 See, e.g., Breton and Wintrobe (1982) at pp. 62–4.
46 Maurer, Fischer, Schwer, Stähler, and Stähler (2009).
47 von Engelhardt and Maurer (2012).
48 Maitland (1985) at p. 133. ("In their attempts to make over managers' value systems and restructure the modern corporation, [corporate responsibility advocates] have largely neglected the very real limits on managers' discretion that result from the operation of a market economy. As a consequence of these limits, managers are largely *unable* to consider their firms' impact on society or to subordinate profit-maximization to social objectives, no matter how well-intentioned they are").
49 See, e.g., Mann (2013) (academic and government estimates place economic value of saving a life "about an order of magnitude" higher than wrongful death cases).
50 For a formal proof, see von Engelhardt and Maurer (2012) at pp. 8–9 and 19.
51 See, e.g., Lenox and Nash (2003); King and Lenox (2000).
52 Lytton and McAllister (2014) at p. 312; cf. Jaffee and Russell (2009).
53 von Engelhardt and Maurer (2012) at p. 1.
54 For a formal proof, see von Engelhardt and Maurer (2012) at pp. 9 and 19–20.
55 For a formal proof, see Id. at p. 10.
56 Vogel (2010) at p. 479 ("Moreover, the public often does not distinguish among the social or environmental practices of firms in the same industry"). In these cases, it is often cheaper to regulate an entire industry than to allocate blame. King, Lenox and Barnett (2002).
57 Vogel (2010) at p. 479.
58 Gunningham and Rees (1997) at p. 394 ("[T]he ability to control free riding increases" where "enterprises are aware of each other's behavior and can detect noncompliance" and "have a history of effective cooperative action (e.g., an existing association)").
59 Vogel (2005) at p. 13 (firm managers balance claims of shareholders, consumers, and the public against personal commitment to civic purposes).
60 Vandenbergh (2006–07) at p. 948.
61 Id.
62 Vogel (2005) at p. 49; Mayer and Gereffi (2010) at pp. 9–10. ("When demand for a product is less a function of observable utility than of constructed brand identity,

firms are more vulnerable to social pressure"); Börzel, Thauer, and Hönke 2011) ("[O]nce a firm starts engaging in self-regulation ... competitors will have strong incentives to follow suit for fear of losing market shares").

63 Steering Committee (2012) at p. 51.
64 Anderton (2000).
65 Bumas (1999).
66 Id.
67 Steering Committee (2012) at p. ES-12 (standards benefit companies through market share).
68 See, e.g., Conroy (2007) at p. 12 (Nike's revenue fell by 16 percent and its stock price by 57 percent following media criticism in 1997–98); Fuchs et al. (2009) at p. 40 (criticism of Kenyan labor and pesticide policies produced significant "market losses for UK retailers"); Borelli (2016) (arguing that Target shares fell 1.5 percent below its nearest competitors after one million consumers signed a petition protesting the firm's transgender bathroom policies in 2016); Anon. (2010) (arguing that Target lost more than $1.3 billion (3.5 percent) in stock market capitalization after 200,000 consumers signed a petition protesting the company's support for a family-values politician); cf. Strassel (2016) at p. 210 (Target apologized and promised to stop spending money on political candidates following 2010 protests); Steering Committee (2012) (finding "anecdotal evidence" of private standards' impact on share price and brand/reputation value.); Pritchett and TIryakian (2017) (reporting that Target CEO's political stands led to "a widespread boycott, and erosion of market share and ... a 40% drop in Target's stock price between April 2016 and July 2017).
69 Maurer (2014).
70 Strassel (2016) at p. 70.
71 Langley (2014) (hackers' 2013 attacks against Target reportedly affected 70 million–110 million customers or about one-third of the U.S. population); Tomasky (2010) (81.5 million U.S. voters cast ballots in the 2010 midterm elections); U.S. Federal Elections Commission (2012) (129.1 million U.S. voters cast ballots in the 2012 presidential election).
72 Vogel (2007) at p. 16. Some observers claim that socially responsible corporations are able to pay lower wages, although the evidence is controversial. Id. at p. 58.
73 Id. at p. 56. Ninety-seven percent of business students claim that they would accept an average of 14 percent lower income to work for companies with "a better reputation for corporate social responsibility and ethics." Id.
74 Id. at pp. 57–8.
75 Strassel (2016) at p 251 (comparing Silicon Valley firms to traditional companies like Pepsi or McDonald's: "These firms are more motivated by their employees, and it is part of the culture that the CEO has coffee with them on Friday to hear and address their concerns").
76 Steering Committee (2012) at p. ES-13 (reporting "anecdotal evidence" that standards improve employee-retention rates).
77 Vandenbergh (2006–07) at p. 948 ("Studies suggest that costs will need to be subsumed within the supply chain for certified goods to be attractive to a broad mass of consumers").
78 Vogel (2007); but see, DeLoecker and Eeckhout (2017) (reporting increased market power of US firms since 1980).
79 Anon. (2008) at pp. 6–8.

80 Anon. (2008) at p. 7.
81 Anon. (2008) at p. 8. But see Vogel (2010) at p. 489 (noting counterexamples in which private standards remain "on the periphery of the firm's business strategies").
82 Vogel (2010) at p. 489.
83 Banjo (2014).
84 Pitofsky (1998). ("[E]nforcement agencies have demonstrated in their case selection and advisory opinions that they are not in the business of challenging legitimate and well-intentioned self-regulation ….").
85 Id. (arguing that privation regulation is the only way to regulate minor matters like "beanball artists" in baseball).

4 Commercial Self-Governance II: Private Politics

1 Williamson (1975).
2 Gunningham and Rees (1997) at p. 373.
3 Vogel (2010) at p. 474.
4 Overdevest and Zeitlin (2014) at p. 33; see also Cashore, Auld, and Newsom (2004); Black (2008).
5 Vogel (2010) at p. 474.
6 Overdevest and Zeitlin (2014) at p. 33; see also Cashore et al. (2004); Black (2008).
7 Meidinger (2007) at p. 517.
8 Steering Committee (2012) at p. 50.
9 Abbott and Snidal (2009) at pp. 561–2.
10 Id.
11 Vogel (2007).
12 Steering Committee (2012) at p. 37.
13 Auld et al. (2008) ("Evolution may occur through competition with other third-party certification systems").
14 Abbott and Snidal (2009).
15 Purnhagen (2014). The downside is that differences between competing standards are often "mainly bureaucratic in nature" leading to "an unnecessary and costly burden". Id.
16 Auld et al. (2008) ("This competition can lead business associations that first initiated voluntary soft codes of conduct to be ultimately spun off as independent organizations").
17 Overdevest and Zeitlin (2014) at p. 23.
18 Eberlein, Abbott, Black, Meidinger, and Wood (2014).
19 Meidinger (2007) at pp. 514, 518.
20 Id. at p. 531.
21 Meidinger (2008); see also Overdevest and Zeitlin (2014).
22 Steering Committee (2012) at p. 37.
23 Meidinger (2007) at p. 519.
24 Steering Committee (2012b) at p. A-185.
25 This is consistent with the older political science orthodoxy that perfect competition and private standards are incompatible. McConnell (1963).
26 Steering Committee (2012) at p. ES-13 (citing "anecdotal evidence").

27 Hicks (1935) at p. 8. For a recent example in which a pharmaceutical executive sold off a lucrative business rather than continue to face accusations of price-gouging, see Joseph Walker, "Drug-Price Revolution Prods a Pioneer to Cash Out," *Wall Street Journal* (May 3 2017).
28 Vogel (2005); Steering Committee (2012a) at p. A-183 (empirical literature suggests that firms are moved by personal, social, and industry norms).
29 Steering Committee (2012(b)) at p. A-183.
30 Williamson (1975).
31 Ostrom (1990) at p. 196 ("The larger the resource system and/or the number of appropriators, and the more unpredictable the flow of resource units and the market prices for these units, the more difficult and costly it is for anyone to obtain accurate information about the condition of the resource itself and the likely value of the flow of resource units under any set of rules").
32 Langley (2014); Wikipedia (n.d.).
33 Reynolds (2006).
34 U.S. Census Bureau (n.d.(a), (b)). See, generally, McElwee (2014) (presenting detailed turnout statistics for 2008, 2010, and 2012).
35 McGinnis (1969); see also Weber (1919).
36 Mayer and Gereffi (2010) at p. 11.
37 Meidinger (2007) at 514–15 and 534.
38 Abbot and Snidal (2009a) at p. 559; Rotherham (2011) (arguing that negative attacks that depress one standard's popularity without raising the other's ultimately weaken both sides).
39 Steering Committee (2012 (a)) at p. 105; Steering Committee (2012(b)) at p. A-189; Abbot and Snidal (2009a) at p. 559.
40 Williamson (1975).
41 Farrell and Saloner (1988).
42 Meidinger (2007) at pp. 531, 514 (The possibility of multiple standards opens the door to "democratic experimentalism"); Abbott and Snidal (2009) at p. 527 ("Competition, demonstration effects, and other interactions help the system as a whole to learn from and scale up successful regulatory experiments … Of course, multiplicity entails transactions costs, but New Governance advocates see the benefits as outweighing them in most cases").
43 Adapted from Maurer (2013).
44 Abbott and Snidal (2009) at p. 554.
45 Id. at p. 557.
46 Meidinger (2007) at p. 531.
47 McCubbins et al. (1984); see also Foreman (1988) (active monitoring tends to be inefficient except where congressional staff have accumulated experience from past investigations).
48 McCubbins et al. (1984).
49 Ostrom (1990) at p. 198. (no attempt will be made to change private rules "If the expected costs of transforming the rules are higher than the net benefits to be gained"). This dynamic is partly offset by the fact that the cost of completing the private standard also falls over time.

5 Legacy: Academic Self-Governance in Modern Times

1. Krimsky (1982) at p. 26.
2. Frederickson (1991) at p. 262 and 265; Lear (1978) at p. 24; Krimsky (1982) at p. 59.
3. Frederickson (1991) at p. 266.
4. Krimsky (1982) at p. 30.
5. Lear (1978) at p. 21. Pollack had been passed over as an assistant professor before coming to Cold Spring Harbor. Id. at p. 23.
6. Krimsky (1982) at pp. 29–30.
7. Lear (1978) at p. 26.
8. Id. at p. 21; Frederickson (1991) at p. 266.
9. Lear (1978) at p. 26.
10. Id. at p. 26.
11. Krimsky (1982) at pp. 31–2.
12. Frederickson (1991) at p. 267.
13. Krimsky (1982) at pp. 31–2.
14. Id. at p. 36.
15. Lear (1978) at p. 37.
16. Id.
17. Krimsky (1982) at pp. 42–3.
18. Lear (1978) at p. 29; Frederickson (1991) at p. 267.
19. Lear (1978) at pp. 32–3; Krimsky (1982) at pp. 33, 42–3; Frederickson (1991) at p. 267.
20. Frederickson (1991) at p. 267.
21. Wade (1977) at p. 32; Lear (1978) at p. 33.
22. Krimsky (1982) at p. 44.
23. Lear (1978) at p. 33; Krimsky (1982) at p. 45.
24. Krimsky (1982) at p. 46.
25. Lear (1978) at pp. 35, 134; Rogers (1977) at p. 70. Some extramural researchers ignored the memorandum's restrictions. Frederickson (1991) at p. 287.
26. Krimsky (1982) at pp. 54–5.
27. Frederickson (1991) at p. 287; Lear (1978) at p. 134. Krimsky dates the NIAID decision from October 1973. Krimsky (1982) at p. 55.
28. Krimsky (1982) at p. 55.
29. Lear (1978) at pp. 65–6.
30. Krimsky (1982) at p. 73.
31. Id.
32. Lear (1978) at p. 71; Krimsky (1982) at p. 74.
33. Rogers (1977) at p. 44; cf. Lear (1978) at p. 71 (Lederberg would almost certainly have refused).
34. Lear (1978) at p. 72.
35. Krimsky (1982) at p. 74; Rogers (1977) at p. 42.
36. Lear (1978) at pp. 72–3; Krimsky (1982) at pp. 75–6.
37. Krimsky (1982) at p. 75 (quoting Maxine Singer); see also Berg and Singer (1995) at p. 9012 (some scientists came to believe "that the public debate was a great threat and that the fallout of claim and counterclaim would bring debilitating restrictions or even prohibitions on molecular biological research").

38 Krimsky (1982) at p. 75.
39 Lear (1978) at p. 83.
40 Id. at pp. 83–4.
41 Id. at p. 84.
42 Rogers (1977) at p. 43.
43 Id. at p. 44.
44 Krimsky (1982) at p. 76.
45 Lear (1978) at pp. 75–6.
46 Krimsky (1982) at p. 80; Lear (1978) at p. 78.
47 Krimsky (1982) at p. 80; Lear (1978) at p. 78.
48 Lear (1978) at pp. 78, 80; Krimsky (1982) at p. 82. This ran contrary to Roblin's initial instinct that the committee should include "someone more radical or skeptical." Id.
49 Lear (1978) at pp. 81, 84–5; Rogers (1977) at p. 45.
50 Frederickson (1991) at p. 272.
51 Rogers (1977) at p. 45.
52 Lear (1978) at p. 82; Rogers (1977) at p. 45; Krimsky (1982) at p. 83.
53 Berg et al. (1974).
54 Krimsky (1982) at p. 84.
55 Rogers (1977) at p. 45.
56 Krimsky (1982) at p. 103.
57 Rogers (1977) at p. 45; Krimsky (1982) at p. 113.
58 Lear (1978) at p. 95.
59 Krimsky (1982) at p. 82.
60 Quoted in Wade (1977) at p. 38; see also Lear (1978) at p. 96.
61 Wade (1977) at p. 38; Rogers (1977) at pp. 47–8.
62 Wade (1977) at pp. 38–9.
63 Lear (1978) at p. 96.
64 Id. at p. 39.
65 Id. at p. 38; Rogers (1977) at pp. 47–8.
66 Rogers (1977) at p. 44.
67 Wade (1977) at pp. 39; see also Berg and Singer (1995) at p. 9011 ("In spite of widespread consternation among many scientists about the proscriptions, the validity of the concerns, and the manner in which they were announced, the moratorium was universally observed"); Berg (2008) ("Scientists around the world hotly debated the wisdom of our call for caution … Yet the moratorium was universally upheld in academic and industrial research centers"); Rogers (1977) at p. 48 ("Yet the moratorium was almost universally observed in the eight months between publication of the letter and … Asilomar").
68 Lear (1978) at pp. 99, 102.
69 Id. at p. 102.
70 Id.
71 Id. at pp. 103, 106; Frederickson (1991) at p. 276.
72 Frederickson (1991) at p. 276.
73 Lear (1978) at p. 96.
74 Id. at p. 37.
75 Frederickson (1991) at p. 268; Krimsky at pp. 62, 64.

76 Frederickson (1991) at pp. 268–9. For the conference proceedings, see Hellman, Oxman, and Pollack (1973).
77 Frederickson (1991) at p. 269.
78 Id. at p. 287.
79 Rogers (1977) at p. 43.
80 Frederickson (1991) at p. 274. An NIH observer also attended. Id.
81 Krimsky (1982) at pp. 103, 109.
82 Id. at p. 103; Frederickson (1991) at p. 274.
83 Krimsky (1982) at p. 104.
84 Id. at p. 106.
85 Id. at pp. 107–8.
86 Id. at pp. 108–10.
87 Wade (1975).
88 Lear (1978) at p. 114.
89 Krimsky (1982) at p. 110.
90 Berg (2008).
91 Lear (1978) at p. 124.
92 Frederickson (1991) at p. 279; Krimsky (1982) at p. 136; Rogers (1977) at p. 72.
93 Krimsky (1982) at p. 138; Frederickson (1991) at p. 280.
94 Wade (1977) at p. 41.
95 Frederickson (1991) at p. 258.
96 Berg (2008).
97 Krimsky (1982) at p. 142.
98 Id. at pp. 142–3; Lear (1978) at p. 131.
99 Frederickson (1991) at p. 275.
100 Rogers (1977) at p. 82.
101 Lear (1978) at p. 126.
102 Krimsky (1982) at p. 136.
103 Berg (2008).
104 Rogers (1977) at pp. 60–1.
105 Krimsky (1982) at p. 130.
106 Quoted in Wade (1977) at pp. 45–6; see also, Berg (2008); Rogers (1977) at p. 65.
107 Krimsky (1982) at p. 131; Frederickson (1991) at p. 279.
108 Lear (1978) at p. 135 (quoting Lewis).
109 Frederickson (1991) at p. 279; Lear (1978) at p. 135 (quoting Lewis).
110 Rogers (1977) at pp. 74–5.
111 Frederickson (1991) at p. 279.
112 Krimsky (1982) at p. 145.
113 Id. at p. 135.
114 Rogers (1977) at pp. 74–5.
115 Id. at p. 86.
116 Id. at pp. 82–3.
117 Lear (1978) at p. 142.
118 Rogers (1977) at p. 83.
119 Id.; Wade (1977) at pp. 50–1. Singer's copy of the organizing committee's draft is available in facsimile at Singer (n.d.).
120 Krimsky (1982) at p. 144.

121 Id. at pp. 143–4.
122 Lear (1978) at p. 142.
123 Krimsky (1982) at pp. 143–4.
124 Wade (1977) at p. 51.
125 Rogers (1977) at p. 84.
126 Id.; Lear (1978) at p. 142.
127 Rogers (1977) at p. 85; Frederickson (1991) at p. 282.
128 Rogers (1977) at p. 85; Lear (1978) at p. 142; Krimsky (1982) at p. 146.
129 Lear (1978) at p. 142. The committee was careful to specify that its recommended precautions could change over time, vary from country to country, or be reduced after inherently safe vectors became available. Id.
130 Rogers (1977) reports that there were "never more than five or six hands" raised in opposition at any point in the proceedings. Id. at p. 96.
131 Id. at pp. 94, 98 (the paragraph on virus work passed "with many abstentions but only a handful of active dissenters").
132 Id. at p. 98.
133 Id.
134 Krimsky (1982) at p. 147; Rogers (1977) at p. 100.
135 Rogers (1977) at p. 100.
136 Frederickson (1991) at pp. 276, 290–1, and n. 44. Anderson explained the ambiguity on the ground that "I hadn't had time to consider all the issues and therefore couldn't be completely negative." Id.
137 Rogers (1977) at p. 100; Frederickson (1991) at p. 282. Wade claims that Cohen and Lederberg were the only dissenters. Wade (1977) at p. 52. Lear claims that Cohen, Lederberg, and Watson were the only ones to vote "no" on the statement as a whole. Lear (1978) at p. 145.
138 Rogers (1977) at pp. 94, 96.
139 Lear (1978) at p. 145.
140 Rogers (1977) at p. 96.
141 Krimsky (1982) at p. 147.
142 Frederickson (1991) at p. 259.
143 Krimsky (1982) at p. 147.
144 Id. at p. 156.
145 Lear (1978) at p. 124; Wade (1977) at p. 56.
146 Krimsky (1982) at p. 182; Wade (1977) at p. 56.
147 Krimsky (1982) at p. 163.
148 Berg (2008). The NIH took the first of several steps to relax the standard at the end of 1978. Frederickson (1991) at p. 283.
149 Krimsky (1982) at p. 199.
150 Except as otherwise noted, this section is based on the account found in Maurer (2006).
151 Wikipedia (n.d.(h)).
152 Wikipedia (n.d.(a)).
153 See Chapter 3.
154 Maurer (2011); see also Service (2006) (academic synthetic biologists started to consult ethicists and launch studies in 2004).
155 Maurer (2011) at p. 1395.

156 Id.
157 Cello, Paul, and Wimmer (2002).
158 Steinbruner and Harris (2003) (citing criticisms by biotech pioneer Craig Venter and medical ethicist Arthur Caplan).
159 Stemerding, de Vried, Walhout, and Van Est (2009) at pp. 155, 157.
160 Relman Committee (2006) at p. viii (arguing that "the future is now" and regulation is needed).
161 Royal Society and Wellcome Trust (2004) at p. 1 ("Self governance by the scientific community [is] favoured, rather than new legislation.").
162 Anon. (2004) (describing how the First International Conference on Synthetic Biology hosted "moderated discussions to help begin to explore ... current and future biological risk"); Carlson (2005) (arguing that SB1.0 members were moved both by the actual threat and by "the potential public backlash it may incite").
163 MIT News Office (2005); Pennisi (2005); Anon. (2005). In the event, most of the options called for conventional regulation on the Asilomar pattern. Garfinkel et al. (2007).
164 Maurer, Lucas, and Terrell (2006) at pp. 2–3, 13–25, and App. A at p. 4.
165 Maurer and Zoloth (2007).
166 The author talked with several conference participants at length shortly after the meeting. See also, Maurer and Zoloth (2007) (members were concerned that the conference needed a constitution before it could vote, or that a vote might be divisive).
167 Service (2006) at p. 1116 (*Science*: SB2.0 "only took baby steps toward self-regulation"); Aldhous (2006) (*New Scientist*: SB2.0 rejected proposals as "controversial" and "too much for synthetic biologists themselves" and premature "until more research had been done on screening and other options").
168 Parens, Johnson, and Moses (2009) at p. 12.
169 Second International Meeting on Synthetic Biology (2006).
170 Kelle (2009) at p. 525 (the declaration promised that "an open working group" would improve the "existing software tools for screening DNA sequences").
171 Maurer and Zoloth (2007).
172 Id.
173 Steinbruner and Harris (2003).
174 Id.
175 Zilinskas and Tucker (2002).
176 Id.
177 Id.
178 Check Hayden (2003).
179 Id.
180 Journal Editors and Authors Group (2003).
181 For a list of names and institutions, see Maurer (2011(a))
182 Journal Editors and Authors Group (2003).
183 See, generally, Steinbruner and Harris (2003).
184 Kennedy (2005).
185 Selgelid (2007) at pp. 35, 41.
186 Anon. (2011).
187 Weaver (2012).

6 Academic Self-Governance: Power and Politics

1. Polanyi (1962 [1942]).
2. Id.
3. Geiger (1984).
4. Id.
5. The UK Royal Society's *Philosophical Transactions* was the first journal to adopt peer review in 1665. Chubin and Hackett (1990) at p. 19.
6. Id. at p. 20 (describing National Cancer Advisory Council).
7. See, e.g., Ozoliņa et al. (2009) at pp. 10–11; see also Chubin and Hackett (1990) at p. 19 (peer review became the default position for allocating research funds in the 1940s and '50s); Frederickson (1991) at p. 260 (postwar science resources "were primarily distributed to individual scientists on the basis of judgments of proposals by their scientific peers, managed on a national basis").
8. Lear (1978) at p. 152.
9. Frederickson (1991) at p. 260.
10. Id.
11. See, e.g., Ozoliņa et al. (2009) at pp. 10–11.
12. Id. at pp. 10–14.
13. Id.
14. Chubin and Hackett (1990) at p. 68; see also Bornmann (2011) at p. 204 (statistical studies suggest that up to 50 percent of the funding decision has no discernible relation to applicant or project characteristics, but instead depends on "apparently random elements which might be characterized as the luck of the reviewer draw").
15. Kreuter (2012).
16. Bornmann (2011) at p. 200.
17. Campbell et al. (2002).
18. Bornmann (2011).
19. For a prominent example involving synthetic biology, see Synberc (2015).
20. Breton and Wintrobe (1982) at p. 63.
21. Cabral at p. 6.
22. Ostrom (1990).
23. Breton and Wintrobe (1982) at p. 160.
24. Cabral (2005) at p. 1.
25. See generally, Jackson (2008).
26. Ostrom (1990) at p. 89 ("Further, a reputation for keeping promises, honest dealings, and reliability in one arena is a valuable asset. Prudent, long-term self-interest reinforces the acceptance of the norms of proper behavior").
27. Cabral (2005) at p. 3. The formal analysis invokes Baysian updating. Id.
28. Id. at pp. 7–8; Green and Porter (1984); Abreu, Pearce, and Stacchetti (1990).
29. Cabral (2005) at pp. 12–13; Shapiro (1983).
30. Scotchmer and Farrell (1988).
31. See, e.g., Pickering (1984) (particle physics collaborations).
32. Rhodes (n.d.).
33. Maurer (2012) (explaining how commercial software development suppresses competition among participating firms).
34. Wikipedia (n.d.(e)).

35 Berger (2012) at p. 22.
36 Id. at p. 13 ("several well-established and formerly successful laboratories have closed because the laboratory heads could no longer secure funding"); Brokenpipeline.org (2008) at pp. 4, 16 (presenting anecdotal evidence of talented scientists who left research for lack of funding); Wikipedia (n.d.(f)) (grant recipients publish more and have better career prospects).
37 See, e.g., Brokenpipeline.org. (2008).
38 The study surveyed a random sample of 126 faculty investigators at ten medical schools who failed to obtain funding for 153 projects in 1970–1. Carter, et al. (1978) at pp. v, 5. All interviews were conducted in 1978. Id. at p. 19.
39 Carter et al. (1978) at p. vi.
40 Bornmann (2011) at p. 200 (noting that effect has grown as internal university funding erodes); Guston (2003).
41 Rutherford and Maurer (2009) at pp. 130–3.
42 Carter et al. (1978) at p. 19.
43 Id. at p. 11. The rate was closer to one-third for rejected NSF and NCI proposals. Chubin and Hackett (1990) at p. 63.
44 Carter et al. (1978) at p. 35.
45 Baringa (2003) at p. 5458 (citing David Baltimore).
46 Berg (2008) ("economic self-interest" would doom a contemporary Asilomar to "acrimony and policy stagnation").
47 Ostrom (1990) at p. 200.
48 Id.
49 Bush (1945).
50 See, e.g., Ozoliṇa et al. (2009) at p. 14.
51 Landeweerd et al. (2015) at p. 7.
52 Id. at p. 8; see also Chilvers and Macnaghten (2011) at p. 25 (describing collaborations with social scientists designed to provide real-time assessment and governance when new technologies are developed).
53 Landeweerd et al. (2015) at p. 8.
54 Id.
55 Id. at p. 16.
56 See, e.g., Ozoliṇa et al. (2009) at p. 28.
57 Landeweerd et al. (2015).
58 Chilvers and Macnaghten (2011) at p. 28 (the UK's formal advisory bodies for biology include lawyers, social scientists, philosophers, ethicists, industry, and civil society organizations).
59 Landeweerd et al. (2015) at p. 12.
60 Id. at p. 14.
61 Id. at p. 13.

7 Legitimacy

1 Meidinger (2007) at p. 140. See also Abbott and Snidal (2009a) at n. 218.
2 Abbott and Snidal (2009a) at n. 219 and citations therein; Koppel (2008) at p. 192 ("there are no universally shared criteria" for evaluating the legitimacy of global governance); Bernstein (2011) at p. 22.

3 Meidinger (2008) at pp. 522–3.
4 Dixon (1978) at p. 17.
5 Russell (2014).
6 See, e.g., Davies and Lynch (2002) at p. 143.
7 Dixon (1978) at pp. 8–9, 17.
8 ISEAL (2013, 2014).
9 Id.
10 Bodansky (2008).
11 Abbott and Snidal (2009a) at p. 555.
12 Bodansky (2008); Grant and Keohane (2005) at p. 33 (democratic accountability requires a "coherent and well-defined global public").
13 Grant and Keohane (2005) at p. 34.
14 Meidinger (2008) at p. 526.
15 Id. at p. 524.
16 Id. at p. 525.
17 Roberts (2012) at p. A-222.
18 Meidinger (2008); Bernstein and Cashore (2007) at p. 353 (reporting criticism that international institutions should be more accountable to broader affected publics).
19 Waldman (2016) at p. 41 (the framers "had less horror of amendments than later Americans did").
20 *Korematsu v. United States*, 323 US 214 (1944) (affirming internment of Japanese-American citizens during World War II).
21 Bernstein (2011) at p. 19.
22 Id.
23 Meidinger (2008) at p. 523. ("Thus, in aggregative models, the people either simply vote on their laws or vote for representatives who then vote on laws. In deliberative models, the people either reason together to develop the best laws, or their representatives do so").
24 Bernstein (2011) at p. 19; Cashore (2002) at p. 510 ("There are no popular elections under [private] governance systems, and no one can be incarcerated or fined for failing to comply"); but see, Scott, Cafaggi, and Senden (2011) at p. 13 ("[I]t is arguable that the focus on electoral politics constitutes a privileging of *one form* of governance which complies with a meta-principle of the right to self-governance, and that other mechanisms of self-governance should be given equal prominence").
25 Vanderburgh (2006/7) at p. 942. (legitimacy arguably requires "democratic accountability to the electorate" and "publicly elected officials"); Balleisen and Eisner (2009) at pp. 129–30 (semiempirical argument that private governance is only "constructive" where the state furnishes "clear missions," maintains "close watch," and exercises a "credible threat" of intervening against private actors that "do not meet their obligations"); Abbott and Snidal (2009a) at pp. 545–6 ("Liberal critics argue that democratic states are the sole legitimate regulators; only they grant each citizen an equal electoral voice and establish clear standards and procedures for representation and accountability. From this perspective, [private governance] can be seen as bypassing and weakening democratic decision making"); Grant and Keohane (2005) at p. 34 (arguing that democracy is poorly defined where no "legal institutions define a public with authority to act globally"); Short (2013) at p. 23

("[S]elf-regulation works best when it is not really self-regulation at all, but when it constitutes regulated organizations as more governable institutions within a robust regulatory regime"); Cafaggi and Renda (2012) at p. 12 ("[F]or any private governance scheme to be viable from a public policy perspective, regardless of whether it is a spontaneous scheme or a policy-induced scheme, suitable monitoring mechanisms must be in place for public policymakers to be able to observe whether the behavior of private regulators is sufficiently aligned with the public interest").

26 Strassel (2016) at p. 227.
27 Meidinger (2008) at p. 527 ("While FSC provides resources and venues so that poor and underrepresented interests can participate in its deliberations, the representatives "are not chosen by constituents" and only "assumed to be able to represent them for others").
28 Meidinger (2008) (arguing that external politics pressures private governance organizations "to respond to or even anticipate public demands"); Eberlein et al. (2014).
29 Dixon (1978) at p. 37.
30 ISEAL (2014). See, e.g., Black (2008) at p. 142.
31 Mattli and Woods (2009) at p. 4 (mechanisms that are not fair, transparent, accessible, and open are subject to capture).
32 Accumulated human capital including experienced social entrepreneurs and shared ideas for change also matter. Mattli and Woods (2009) at p. 4.
33 Meidinger (2008); Scott et al. (2011) at pp. 13, 16 (competing networks "may be as important sources of legitimacy as the ballot box").
34 Abbott and Snidal (2009a) at p. 557; Mattli and Woods (2009) at p. 2.
35 Auld et al. (2008) at p. 423.
36 Scott et al. (2011) at pp. 2, 4.
37 Bodansky (2008); Black (2008) at p. 142.
38 Abbott and Snidal (2009a) at p. 529. See also, Steering Committee (2012) at p. 39 (Private governance may also provide more "effective mechanisms for articulating goals, testing them in practice, and revising them.")
39 Id. at p. 60.
40 Dixon (1978) at p. 28.
41 Lytton (2014) at pp. 540, 564. Lytton remarks that, while government could theoretically support regulation on a fee-for-service basis, this faces "stiff political resistance." Id. at p. 565.
42 This is not true where government seeks recoupment. Roberts (2012) at p. A-244.
43 Lytton (2014) at p. 566.
44 Id. at p. 565.
45 Mayer and Gereffi (2010) at p. 1 (claiming that private governance "is best seen as a response to ... economic globalization and the inadequacy of public governance institutions in addressing them").
46 Abbott and Snidal (2009a) at pp. 558–9 (noting that even democratic states fail to "take account of interests that are widely distributed internationally" and that "independent decisions by multiple states ... may be inconsistent").
47 Mayer and Gereffi (2010) at p. 4.
48 Abbott and Snidal (2009a) at p. 527–8; Bodansky (2008) (scholars have turned to deliberative theories in frustration with the special problems of transnational

democracy); Meidinger (2008) at p. 529 ("Deliberative democracy is sometimes suggested as a way to avoid representation and aggregation objections").
49 Black (2008) at p. 142.
50 Id. at p. 146.
51 Rose (2016) (arguing that the right to speech is not instrumental but inherent in our human dignity and our status as moral beings).
52 Steering Committee (2012) at pp. 37–8.
53 Existential philosophers famously claim that even useless acts can supply meaning to our lives. Cf. Camus (1955 [1942]).
54 Abbott and Snidal (2009a) at p. 555.
55 Nelson (2015); Chait (2012).
56 *United States v. Aluminum Co. of America*, 148 F.2d 416 (2d Cir. 1945).
57 Dixon (1978) at p. 2 (private industrial standards' claim to acceptance "is predicated on a broad consensus based on careful study by persons specially informed or affected").
58 Id. at p. 48 ("Well-run committees can anticipate areas of unhappiness and seek out input from those affected. Failing that, openness, procedural fairness, and appeal procedures can stand in").
59 See, e.g., Davies and Lynch (2002) at p. 143.
60 Cafaggi and Renda (2012) at p. 4 (some lawyers see private self-regulation "as an example of modern corporatism").
61 Meidinger (2008) at p. 529; see also Mattli and Woods (2009) at p. 43 (mechanisms that are *not* fair, transparent, and accessible prevent parties from "participat[ing] meaningfully" and are illegitimate).
62 Scott et al. (2011) at p. 13.
63 Dixon (1978) at p. 50.
64 Meidinger (2008) at p. 529.
65 Schleifer and Bloomfield (2015) at p. 2; Bodansky (2008); Suchman (1995) at p. 580.
66 Ponte, Gibbon, and Vestergaard (2011) at p. 17 (output legitimacy is not just about winning in the marketplace, but also striking the right balance of costs and benefits); Kjaer (2004) at p. 12; Black (2008) at p. 142.
67 Black (2008) at p. 146; cf. Suchman (1995) at p. 579 (moral legitimacy is "based on perception of society's best interest").
68 Black (2008) at p. 146.
69 Suchman (1995) at p. 578; Schleifer and Bloomfield (2015) at p. 2.
70 Suchman (1995) at pp. 582–3.
71 Cashore, Auld, and Newsom (2003) at p. 228; Schleifer and Bloomfield (2015) at p. 2.
72 Scott et al. (2011) at p. 14.
73 Bodansky (2008); See also, Bernstein (2011) at p. 20 (collecting literature that treats political legitimacy as "the acceptance and justification of shared rule by a community"); Suchman (1995) (sociology and organizational theory see legitimacy as a "relational concept" that is "granted or denied by an institution's audiences").
74 Bodansky (2008) at p. 20.
75 Schleifer and Bloomfield (2015) at p. 2.

76 Bernstein (2011); Bernstein and Cashore (2007).
77 Black (2008) at p. 142.
78 Schleifer and Bloomfield (2015) at p. 2 ("we find no instances of this type of legitimacy in our cases and, therefore, leave this dimension aside for the purposes of our study"); but see Scott et al. (2011) at p. 14 (arguing that traditional industry standard-setting rules have produced "an increasing inability to imagine other ways in which such standards might be set").
79 Abbott and Snidal (2009a) at p. 555 (FSC and other organizations have pragmatically adopted deliberation and other democratic virtues even with "convincing democratic theories for the global sphere ... lacking").
80 Scott et al. (2011) at p. 13 ("[T]he potential for judicial accountability may be as important sources of legitimacy as the ballot box"). The authors add that government endorsement is more valuable when private standards compete, so that governments are able to give recognition to some but not others. Id. at p. 18; see also Cafaggi and Renda (2012) at p. 13.
81 Hobson (2009) at p. 632; Bernstein (2011) at pp. 22–3.
82 Bodansky (2008).
83 Suchman (1995) at p. 581.
84 But see Waldman (2016) at p. 31 (quoting Alexander Hamilton, "As [groups] are a collection of individual men, which ought we to respect most, the rights of the people composing them, or the artificial beings that result from composition").
85 Scott et al. (2011) at pp. 5–6 (arguing that promulgation of norms is a valuable activity regardless of "whether they are publicly or privately promulgated").
86 But see Chubin and Hackett (1990) at p. 29 (critics charge that peer review takes power away from elected officials, lets scientists ignore social needs, discriminates against researchers at low-prestige and nonacademic institutions, and protects academic proposals of doubtful value).

8 Law

1 Kolk (2013) (4C collaboration believed that any explicit discussion of coffee prices would violate competitions policy); Potts (2004) at pp. 13–14; Morris, Seward, and Kurzrok (2012) (arguing that private, voluntary restrictions on nuclear-technology sales do not violate the Sherman Act).
2 Auld et al. (2002) at p. 280 (U.S. Certified Products Association believed that requiring members to purchase FSC lumber would violate the Sherman Act); Künkel et al. (2008) (some producer countries have charged that competition policy concerns are an excuse to avoid price negotiations); Steering Committee (2012) at p. 32 ("To minimize concerns about anti-competitive behavior, roundtables have established antitrust policies that explicitly state the topics that can and cannot be discussed").
3 Ponsoldt (1981) at p. 1 and n. 1; Leary (2004) ("[I]t is unlikely that we will interfere with appropriately tailored industry efforts to restrict" the sale of "legal, but potentially harmful products to children"); Pitofsky (1998) ("Government resources are limited ... Thus, many government agencies, like the FTC, have sought to leverage their limited resources by promoting and encouraging self-regulation").

4 Maitland (1985) (antitrust courts have struck down self-governance schemes designed to, inter alia, eliminate inferior products, impose ethics codes, establish oil tanker safety ratings, and coordinate flights into overcrowded airports).
5 *Hatley v. Am. Quarter Horse Ass'n*, 552 F.2d 646, 655 (5th Cir. 1977).
6 Anon. (1963) at pp. 1000–1.
7 *Hatley*, 522 F.2d at p. 655 (association policy cannot conflict with "good morals," the "bounds of reason," "common sense," "fairness," state law, or "public policy"); *Hennessey v. NCAA*, 564 F.2d 1136, 1144 (5th Cir. 1977) (rules need not be "perfectly suited" or "perfectly fair," so long as they have an "articulable rationale" and are "rationally related to that objective, and free from arbitrariness and invidious discrimination"). For some rare instances in which judges were willing to overrule private bodies, see *James v. Marinship Corp*, 155 P.2d at p. 342 (Cal. 1944) (overturning union rule against admitting black members); *Carroll v. Int'l Bd. of Elec. Workers*, 31 A.2d 223, 225 (N.J. Ch. 1943) (public policy required unions to be "democratic" and admit all who are "reasonably qualified for their trade").
8 Anon. (1963) at p. 997 (social policy is only invoked in the "clearest cases" and where private rules would require members to "commit crimes" or prevent them "from performing public duties such as serving on juries"); *Charles O. Finley & Co. v. Kuhn*, 569 F.2d 527, 544 (7th Cir. 1978) (rules cannot conflict with "the law of the land"); *Falcone v. Middlesex Cnty. Med. Soc'y*, 170 A.2d 791, 800 (N.J. 1961) (striking down actions that "run[] strongly counter to the public policy of our State and the true interests of justice"); *James*, 155 P.2d 329, 337 (Cal. 1944) (actions cannot be contrary to "public policy" or conflict with state or federal law or, in some states, common law precedents where "the law has definitely set its face"); *Jackson v. Am. Yorkshire Club*, 340 F. Supp. 628, 632 (N.D. Iowa 1971) (private rules must be "reasonable," "pursuant to the rules and laws of the organization," made "in good faith," and not in "violation of the law of the land").
9 *Gilder v. PGA Tour, Inc.*, 936 F.2d 417, 424 (9th Cir. 1991) (directors cannot change existing rules to achieve a specific result in a pending matter); *Charles O. Finley & Co. v. Kuhn*, 569 F.2d 527, 544 (7th Cir. 1978) (organization cannot adopt rules in "disregard" of its own "charter or bylaws"); *Hennessey v. NCAA*, 564 F.2d 1136, 1147 (5th Cir. 1977) (rulemaking satisfied due process where the organization followed its own procedures, conducted careful study and deliberation, gave notice to affected parties, and held majority vote). A fortiori, rules adopted by supermajorities command more deference. Id. at p. 1147 (noting that bylaw was changed by "greater than majority vote").
10 *Lindemann v. Am. Horse Shows Ass'n*, 624 N.Y.S. 2d 723, 727 (1994) ("pro forma," "arbitrary," "capricious," or "pretextual" hearings are inadequate); *Hatley*, 552 F.2d at pp. 654–5 (organizations must hold hearings where enforcement requires subjective judgments or discretion); *Jackson*, 340 F. Supp. 628, 632 (N.D. Iowa 1971) (principles of "natural justice" require notice, hearing, and an opportunity to defend); *Charles O. Finley*, 569 F.2d 527 at p. 540 (enforcement must be brought in good faith and avoid "abrupt departures" from prior practice); *STP Corp. v. U.S. Auto Club*, 286 F. Supp. 146, 170 (S.D. Ind. 1968) (private rules enforcement must follow "rudimentary due process procedures" and any actions must be "reasonable, done in good faith, and … not discriminatory"); *Gilder*, 936 F.2d at p. 424 (board of directors could not act for "private gain").

Notes to Pages 167–170   259

11 *Falcone*, 170 A.2d 791, 799 (N.J. 1961).
12 *Marjorie Webster Junior Coll., Inc. v. Middle States Assn. of Colleges & Secondary Schools, Inc.*, 432 F.2d 650 (D.C. Cir. 1970) at p. 658.
13 Anon. (1963) at pp. 1046–7 (courts are "less sympathetic" where there is an "apparent self-serving interest in the standards and the lack of a well-established tradition of self-policing").
14 Courts also accept lower standards where more burdensome procedures could deprive organizations of "the ability to promote ... specific social goals generally believed to be desirable." Anon. (1963) at p. 991.
15 See *Restatement of Torts* § 765 (1939); *Restatement (Second) of Torts* §§ 767, 768 (1977).
16 Anon. (1963) at p. 1068.
17 Verkuil (2006) at p. 433.
18 Id.
19 But see, *Marjorie Webster*, 432 F.2d 650 (D.C. Cir. 1970) at p. 658 (court would not interfere where plaintiff had sufficient resources to withdraw from college accreditation body and establish a competing organization).
20 Maitland (1985) at p. 137.
21 Hofstadter (1991) [1964] at p. 28; Millon (1991 [1988]) at p. 86.
22 Hofstadter (1991) [1964] at p. 28.
23 Burton (1906) at p. 359.
24 Id.
25 Id.
26 *United States v. Trans-Mo. Freight Assn.*, 166 U.S. 290 (1897) at p. 324.
27 *Standard Oil Co. of New Jersey v. United States*, 221 U.S. 1 (1911).
28 Millon (1991) [1988] at pp. 86–7.
29 Davies and Lynch (2002).
30 *Schechter Poultry Corp. v. United States*, 295 U.S. 495 (1935).
31 *Fashion Originators' Guild, Inc. v. FTC*, 312 U.S. 457 (1941).
32 Id. The Court specifically found that FOGA had violated the Sherman Act by narrowing the outlets that retailers could buy from, and requiring guild members to disclose the "intimate details of their individual affairs." Id. at p. 465. The Court added that FOGA had equally violated the Clayton Act, which prohibits contracts that condition the sale of goods on promises not to use or sell competing products. Id. at p. 464.
33 Id. (quoting *Addyston Pipe & Steel Co. v. United States*, 175 U.S. 211, 242 (1899)).
34 Ponsoldt (1981) at p. 25 ("This was arguably an indication by the Court of its intolerance for industry self-regulatory schemes of any kind").
35 Anon. (1963) at p. 1047.
36 *United States v. National Association of Broadcasters*, 536 F. Supp. 149, 164 (D.D.C. 1982) (rules that are voluntary in both theory and actual practice are not subject to the antitrust laws even when backed by the threat of public criticism); *Costello Publishing Co. v Rotelle*, 670 F.2d 1035, 1047 (D.C. Cir. 1981) (lower court should consider whether the Catholic Church's condemnation of a book was an illegal boycott so long as retailers could freely decide not to stock it); *Tropic Film Corp. v. Paramount Pictures Corp.*, 319 F. Supp. (S.D.N.Y. 1970) at p. 1250 (noting that newspapers' decision to follow private rating organization's recommendation not

to advertise adult movies was "entirely voluntary" and that the same result might have equally have resulted from a series of individual judgments based on content alone); *Law v. NCAA*, 134 F.3d 1010, 1020 (10th Cir. 1998) (coercion did not exist where organization lacked significant market power); *Brant v. U.S. Polo Ass'n*, 631 F. Supp. 71, 77 (S.D. Fla. 1986) (suspension was too short to have any "palpable anticompetitive effect" on any relevant market). A more doubtful version of the rule argued that associations could coerce their own members but not nonmembers. *College Athletic Placement Services, Inc. v. NCAA*, 1974 U.S. Dist. LEXIS 7050, at pp. 8–10 (D.N.J. Aug. 22, 1974) ("The principle of the group boycott cases – that it is prima facie unreasonable for a dominant group to combine to coerce" did not apply where challenged restriction was designed to regulate member behavior and impact on third parties was "at best indirect").

37 Elhauge (1991) at p. 742 (arguing that *FOGA* is currently limited to cases where actors controlling "extra-governmental agenc[ies]" have "financial interests that seem[] certain to bias their extra-governmental conduct").
38 Balleisen and Eisner (2009) at pp. 130–1.
39 15 U.S. Code § 1.
40 See, e.g., Federal Trade Commission and Department of Justice (2000).
41 *United States v. Oregon State Medical Society*, 343 U.S. 326 at p. 336 (1952) (emphasis added) (professional society did not violate antitrust laws by issuing ethics rule that barred members from replacing traditional doctor–patient relationship with "tripartite" arrangement that gave employers and insurance companies the right to influence treatment).
42 See, e.g., Kirby and Weymouth (1985); *Marjorie Webster*, 432 F.2d 650 at pp. 654–5 (D.C. Cir. 1970) ("It is possible to conceive of restrictions on eligibility for accreditation that could have little other than a commercial motive … Absent such motives, however, the process of accreditation is an activity distinct from the sphere of commerce; it goes rather to the heart of the concept of education itself").
43 *Marjorie Webster*, 432 F.2d 650 (D.C. Cir. 1970) at p. 654 ("incidental restraint of trade" by organizations dedicated to "the liberal arts and the learned professions" would not trigger antitrust liability "absent a purpose to affect commercial aspects of the profession"); *Cooney v. Am. Horse Shows Ass'n*, 495 F. Supp. 424, at p. 431 (S.D.N.Y. 1980) (disciplinary action against owners accused of doping horses was not anticompetitive unless "the restraint is broader than reasonably necessary to accomplish the legitimate goal of the regulation"); *Brenner v. World Boxing Council*, 675 F.2d 445, at p. 456 (2d Cir. 1982) (sanctions against promoter were not anticompetitive because they "reasonably related to a policy justifying self-regulation"); *Rooffire Alarm Co. v. Royal Indem. Co.*, 202 F. Supp. 166, at p. 169 (E.D. Tenn. 1962) ("An association formed to foster high standards, to mitigate evils in trade existing through lack of knowledge or information, and to encourage fair competitive opportunities" did not violate Sherman Act "merely because it may effect a change in market conditions").
44 *Goldfarb v. Virginia. State Bar*, 421 U.S. 773, 791 (1975).
45 Id. at p. 788; Seib (1985).
46 *National Society of Professional Engineers v. United States*, 435 U.S. at p. 699 (Blackmun, J., concurring).

47 *Community Communications Co., Inc. v. City of Boulder*, 455 U.S. 40, 66 (1982).
48 *Continental T.V., Inc. v. GTE Sylvania, Inc.*, 433 U.S. 36, 53 n.21 (1977).
49 Kirby & Weymouth (1985) at p. 32.
50 *NCAA v. Bd. of Regents of the Univ. of Okla.*, 468 U.S. 85, 101–2 (1984).
51 Id. at p. 102. Two justices dissented, arguing that antitrust law should allow colleges to pursue "legitimate noneconomic goals" even when this impacts "the free market." *NCAA*, 468 U.S. at p. 134.
52 Ponsoldt (1981) ("Safety and other social policies not supported by legislation seem to have been rejected both as defenses in a Rule of Reason case and as arguments against the use of the per se rule"); Seib (1985) at p. 725. ("The judiciary could not defer to what it might consider a higher value" even where the restraint on competition is "insignificant").
53 *Denver Rockets v. All-Pro Mgmt., Inc.*, 325 F. Supp. 1049, 1065, 1066 (C.D. Cal. 1971) (rejecting eligibility rule that required professional basketball players to complete college: "However commendable this desire may be, this court is not in a position to say that this consideration should override the objective of fostering economic competition which is embodied in the antitrust laws. If such a determination is to be made, it must be made by Congress and not the courts"); *Law v. NCAA*, 134 F.3d 1010, 1021–2 (10th Cir. 1998) ("[W]e may not consider such [social] values unless they impact upon competition").
54 Pitofsky (1998) ("[S]ome have cited [*Professional Engineers*], which can be interpreted to rule out consideration of non-economic goals ... in a rule of reason analysis, as a threat to industry self-regulation").
55 See, e.g., *Banks v. NCAA*, 977 F.2d 1081, 1091 (7th Cir. 1992) (Alternate holding: NCAA rules provide a bulwark against the "quick buck" attitudes of professional sports, preserve a "proper focus" on education, and prevent colleges from being transformed into "'minor league farm systems'" for the NFL); *McCormack v. NCAA*, 845 F.2d 1338, 1343 (5th Cir. 1988) (student eligibility rules that serve "purely or primarily noncommercial objectives" held outside antitrust laws); see also Leary (2004) (FTC chairman: broadcast self-censorship "reflects a noncommercial value that ... the courts would uphold"); but see *Clarett v. NFL*, 306 F. Supp. 2d 379 (S.D.N.Y. 2004) at p. 408 (NFL rule designed to protect younger athletes from injury and overtraining, though "laudable," could be "dismissed out of hand" since it had "nothing to do with promoting competition").
56 *Clarett* at p. 409 (questioning whether NFL rule protecting young players could be justified as protecting the NFL brand); cf. *Brenner* (sports leagues can adopt rules for "reasonable objectives" not limited to participant parity, safety, and league integrity); *American Federation of Television & Radio Artists v. National Assn. of Broadcasters*, 407 F. Supp. 900, 901–2 (S.D.N.Y. 1976) (private standard was a "positive response" to criticism by U.S. senator, FCC, and private activists).
57 See, e.g., *NCAA*, 468 U.S. at pp. 101–2 (justifying college recruiting restrictions: "And the integrity of the 'product' cannot be preserved by mutual agreement; if an institution adopted such restrictions unilaterally, its effectiveness as a competitor on the playing field might soon be destroyed").
58 Scott et al. (2011) at p. 6; Maher (2011) at pp. 132–3.
59 Leary (2004) ("When you have a compelling social concern, when the alternative to private regulation may be even more heavy-handed government regulation,

when you are actually asking your members to do something that is against their immediate economic interest – not in aid of it – I think there is a narrow window for consideration of non-economic values in trade association codes and standards").
60 Maurer and Scotchmer (2014).
61 Leibowitz (2005) (remarks by sitting FTC commissioner: "If a board is more than simply a group of competitors, but includes others who would be harmed by an anticompetitive agreement then that makes it less likely that a particular rule is a result of a private agreement to restrict competition").
62 *Silver v. N.Y. Stock Exch.*, 373 U.S. 341, 358 (1963).
63 See id.
64 See, e.g., *Broadcast Music, Inc. v. CBS*, 441 U.S. 1 (1979). (*Silver* applies wherever private rules are necessary to create a market, promote competition, or increase output). For a thorough survey of how courts applied *Silver*, see *Denver Rockets v. All-Pro Management, Inc.*, 325 F. Supp. 1049, 1065 (C.D. Cal. 1971).
65 *Eliason Corp. v. National Sanitation Foundation*, 614 F.2d 126, 130 (6th Cir. 1980) (rules should be periodically revised to reflect new technologies); *Denver Rockets*, 325 F. Supp. at p. 1066 (rule that required professional players to go to college where some did "not desire" or even have "the mental and financial ability to do so" was overbroad); *M&H Tire Co.*, 733 F.2d at p. 984 (rule making must be open to the entire industry and "all interested companies").
66 *Justice v. NCAA*, 577 F. Supp. 356, 381 (D. Ariz. 1983); *Silver v. N.Y. Stock Exch.*, 373 U.S. at pp. 341, 363 n. 15 (1963) (procedural protection for non-members would not "burden" or slow the exchange under the circumstances because it already operated similar protections for members); *Jackson*, 340 F. Supp. at p. 635-6 (in considering decision that would affect member's livelihood, "[t]he directors must, of course, also consider the interests of [their organization]").
67 *Silver*, 373 U.S. at p. 361.
68 *Brenner*, 675 F.2d at p. 456 (restraint was unreasonable where suspension was effected to prevent appellant from engaging in his trade); *McCreery*, 379 F. Supp. at p. 1019 ("The complete control of the registered purebred Black Aberdeen Angus breeding business by the Association means in effect that the Plaintiffs are completely out of the business of raising and selling purebred registered Black Aberdeen Angus cattle while they are under indefinite suspension."); *Jackson*, 340 F. Supp. at p. 635-36 (noting that member's "livelihood as a Yorkshire breeder is dependent upon membership"); *Clarett* at p. 406. (Sherman Act forbids any "contract which unreasonably forbids any one to practice his calling").
69 See, e.g., *Jackson*, 340 F. Supp. at p. 635 (review was limited to "the fairness of the hearing itself" and in "no event would the Court substitute its judgment ... on the merits of the case).

9 Policy and Practice

1 Cafaggi and Renda (2012) at p. 12.
2 Abbott and Snidal (2009b) at p. 58.

3 See Pitofsky (1998) ("Finally, government resources are limited and unlikely to grow in the future. Thus, many government agencies, like the FTC, have sought to leverage their limited resources by promoting and encouraging self-regulation"); Balleisen and Eisner (2009) at p. 145 (private governance "can extend the reach of regulation to areas that are simply beyond the analytical and budgetary capabilities of public regulators").

4 ACUS (1978) at pp. 1–2; Sunstein (1999) ("Government regulation is often the response to market failure, but a code might be better, especially because of its comparative flexibility and because of the informational advantages of private enforcers ... [D]irect regulation may lead the industry to provide benefits more crudely and expensively than if a code were in place").

5 Pitofsky (1998) ("[S]elf-regulation can bring the accumulated judgment and experience of an industry to bear on issues that are sometimes difficult for the government to define with bright line rules"); Leibowitz (2005) ("[S]elf-regulation allows those who know the industry best to help set the rules of the game").

6 Pitofski (1998); Lytton (2014) at p. 564.

7 Abbott and Snidal (2009a) at pp. 528–9.

8 Steering Committee (2012) at p. 74; Webb (2004) at p. 16 (codes test which solutions do and do not work).

9 Büthe and Mattli (2011) at pp. 25 ("Private governance has been encouraged by the "excruciatingly slow pace of standards production" by state bodies "and, in some cases, lack of the technical expertise and financial resources"), 5 (noting how the European Union delegated manufactured goods standards to the private sector because official negotiations were taking too long).

10 Büthe (2010) at pp. 5–6 (describing decade-long delays in negotiating international standards for pesticides, financial reporting, and manufacturing); Mayer and Gereffi (2010) at pp. 1–2 (international regulatory standards are generally weak and hard to enforce).

11 Mayer and Gereffi (2010) at p. 4. (developing governments "lacked the ability, and to some extent the will, to regulate production").

12 Büthe and Mattli (2011); Meyer and Gereffi (2010).

13 See, e.g., Moe (1998) (noting that regulatory agencies are deliberately structured to lock in "a separately conceived and orchestrated political product, fashioned by a unique coalition of legislators and interest groups and designed to promote a particular set of interests").

14 Abbott and Snidal (2009a); Bartley (2011) at pp. 520–1 (slowness and inertia of government and intergovernmental negotiations compared to private forest standards); Pitofsky (1998) (slowness of public compared to private bodies). The difference is particularly dramatic in our artificial-DNA example, where the IASB was able to develop strong, comprehensive standards in eighteen months compared to a federal effort that took more than a decade to produce what turned out to be remarkably limited, nonbinding guidelines.

15 Lack of experimentation is particularly evident in biosecurity, where practically all of today's synthetic-biology proposals repeat or else rediscover ideas that were originally voiced in the first few years after 9/11. For a detailed literature review, see Maurer (2011).

16 Bernstein and Cashore (2007) (arguing that New Self-Governance initiatives emerged "In the absence of effective national and intergovernmental regulation to ameliorate global environmental and social problems"); Vogel (2010) at p. 476 (non-state governance responds to the "inability or unwillingness of states" to regulate global markets and situations).

17 Abbott and Snidal (2009b) at p. 57 (private governance offers an alternative to captured or corrupt governments); Bartley (2011) (self-governance offers a chance to start over outside "old power struggles and structures"); Steering Committee (2012) at p. 88. ("[C]ertification schemes can also *improve regulatory* systems that are fractured, missing, or unable to meet their goals in their current forms. This appears to be a problem especially in international contexts"); Bartley (2011) (private governance is often a response to corruption).

18 Vogel (2010) at pp. 477 ("[M]any NGOs have been repeatedly frustrated by their inability to strengthen treaties … [F]or global activists, lobbying corporations has come to represent a viable, though clearly second-best, alternative to pressing for changes in public policies"), 478 (private action is typically pursued "where there is little prospect of additional regulations being enacted, let alone enforced, especially at the global level or by developing countries").

19 Cafaggi and Renda (2012) at p. 5; Büthe (2010) (private governance offers a second chance when conventional politics deadlocks); Abbott and Snidal (2009a) at p. 501 (failure of treaty and intergovernmental organizations). Our examples include multiple private initiatives that were only organized after government failed to accomplish conventional tasks like funding mutations data, supporting coffee prices, or promulgating forestry and fishing standards.

20 Bartley (2011) at pp. 520–1 (private initiatives avoid the "high costs of intergovernmental coordination"); Abbott and Snidal (2009b) at p. 59 (interstate negotiations are costly and contentious); Cafaggi and Renda (2012) at p. 5 (deadlock is particularly likely in international context).

21 Abbott and Snidal (2009b) at p. 58.

22 Id.

23 Meidinger (2008); Steering Committee (2012) at p. 104 ("Governments should recognize and leverage the fact that standards and certification systems create coalitions of support around areas of policy and regulation that can make it easier for governments to regulate areas that may have been more contentious previously").

24 Steering Committee (2012) at p. 104; see also id. at pp. 85 ("Another key advantage is that the development of standards and certification systems can allow stakeholders to coalesce around a limited number of key issues." This defines the scope of the problem and identifies "a subset of potential actions that already enjoys support from key actors"), 88 (private standards provide "venues for dialogue" and "altered problem definitions"); Webb (2004) at p. 16 (voluntary codes let NGOs, firms, and industry associations engage in "norm conversations" directly with various publics encouraging norm development and implementation).

25 U.S. efforts to enforce China's treaty commitments to intellectual property, for example, remain notoriously incomplete. See, e.g., Jones (2011).

26 Abbott and Snidal (2009a) at p. 528–9; cf. Cafaggi and Renda (2012) at p. 12 (arguing that private regulation can achieve "greater compliance").
27 Abbott and Snidal (2009a) at p. 528–9 ("Collaboration is more conducive than Old Governance procedures to information sharing and learning, an important benefit given the bounded rationality of state regulators. New Governance allows the state to work with regulatory targets to tailor policies to their specific needs and local conditions, rather than forcing uniform rules on disparate circumstances. This creates incentives for firms to exceed mandated standards, and reduces the state's own costs of monitoring and enforcement"); see also Webb (2004) at pp. 16 ("While laws are imposed in a top-down fashion, have a coercive base and are the product of centralized rule systems, voluntary codes are inherently bottom-up, cooperative and consent-based, working only when agreed to by the appropriate attentive publics"); Cafaggi and Renda (2012) at p. 11 ("Private governance focuses on facilitating the interaction of private players as opposed to command and control").
28 Webb (2004) at p. 16.
29 Vogel (2010).
30 Abbott and Snidal (2009a) at p. 559.
31 Id.
32 Id. at p. 521.
33 Gulbrandsen (2014) at p. 76.
34 Künkel et al. (2008).
35 Vanderburgh (2006/7) at p. 968; see also Gulbrandsen (2014) at p. 76. (governments can provide expertise and technical advice); Abbot and Snidal (2009a) at p. 559 (governments can "more systematically encourage learning across the system and disseminate, replicate, and scale up the most successful innovations").
36 Mayer and Gereffi (2010) at pp. 17–18. For example, biology judgments made in the course of screening routine orders could potentially be repurposed to address new science questions and technologies. See Chapter 10.
37 Maurer and Scotchmer (2004).
38 ACUS (1978) at pp 4–9 (recommending that agencies refer matters to private technical committees before drafting mandatory regulation. Formal regulations should defer to or else incorporate private rules wherever possible).
39 Abbot and Snidal (2009a) at p. 521. ("In New Governance, the state remains a significant player, but as an orchestrator rather than a top-down commander").
40 ACUS (1978) at pp. 3–4 (agencies should send nonvoting members to serve on technical committees on the understanding that the agency will not be bound), 5 ("The relationship between the agency and the technical committee should be a cooperative one, and the agency should not seek to dominate the committee").
41 Abbot and Snidal (2009a) at p. 521; Steering Committee (2012) at p. 88 (private standards can provide a laboratory for learning about and demonstrating solutions, which government can, but need not embrace later).
42 Büthe (2010).
43 Id.
44 Abbott and Snidal (2009a) at p. 526.
45 Traditional U.S. domestic practice argues that government should adopt private standards based on "voluntary consensus," "open and regular procedures, including

a process for considering ... negative comments," and rules that favor broad and "balanced" memberships "in an effort to assure representation of varying points of view and avoidance of domination by a single interest." Administrative Conference of the United States (1978).
46 Abbott and Snidal (2009a) at p. 595. The authors argue that this would, among other things, "reduce the risk of capture." Id.
47 Id.
48 Steering Committee (2012) at p. 105. Some nonprofits and consumers publish lists of private standards that they consider acceptable. Roberts (2012) at pp. A-251–2; Gunningham and Rees (1997) at p. 401 ("[I]t may be appropriate for a state agency to endorse a particular self-regulatory program by permitting it to use the agency's logo or other official seal of approval, thereby giving the public greater confidence in the code's credibility").
49 Abbot and Snidal (2009a) at p. 559.
50 Gulbrandsen at p. 76.

10 Extending the Model

1 See, e.g., El Baradei (2006) (International Atomic Energy Agency director: "Either we begin finding creative, outside-the-box solutions or the international nuclear safeguards regime will become obsolete"); Luongo and Williams (2007) at p. 459 (government should "tap[] the power of market-based mechanisms to address ... proliferation challenges"); Noble (2013) (praising private synthetic-biology standards as a model for future self-regulation); Tucker (2010) (arguing that private self-regulation by suppliers and perhaps consumers would be "a better approach"); Carnegie Fund (2012) at p. 10 (arguing that high-profile civilian self-governance models should be extended to nuclear data).
2 Sections 10.2.1 and 10.2.2 owe a large debt to Gretchen Hund and her coworkers, who have spent years studying private nuclear-supply chains. Any errors are mine alone.
3 Maurer (2007).
4 Hund and Seward (2008) at p. A-2; Hund and Elkhamri (2005).
5 Carnegie Fund at p. 6.
6 Albright (2010) at pp. 47–56
7 Carnegie Fund (2012) at p. 6; see also Ottolenghi (2013) at p. 3 (Dusseldorf hosts "thousands of medium and small-sized, family-run companies [that] invent, assemble, produce, and sell some of the best machinery and industrial products in the world").
8 Albright (2010) at pp. 159, 175, 186 (describing Pakistan's and Iran's continued need for spare parts to keep older machine tools running).
9 Id. at p. 180.
10 Wirtz (2010) at p. 274 (front companies tend to be few in number and often persist long after some orders are refused); Ottolenghi (2013) at p. 2 (complexity associated with front companies "must cause endless headaches for the Iranians"); Albright (2010) at pp. 180, 187, 188, 190 (repeated purchase attempts by front companies); Ottolenghi (2013) at pp. 3, 8. (front companies often only last a few months or are

based in managers' homes); Carnegie Fund (2012) (Khan's front companies were little more than a fax machine and an empty office); but see Ottolenghi (2013) at p. 5 (describing Iran's elaborate and expensive front companies).
11 Albright (2010) at p. 92; Fitzpatrick (2007) at p. 43.
12 Wirtz (2010) at p. 275; Carnegie Fund (2012) at pp. 10–11; Albright (2010) at pp. 180, 187.
13 Hund and Seward (2008) at p. 5.
14 Id. at p. 17.
15 Wirtz (2010) at p. 279.
16 Joyner (2004) at p. 109.
17 Carnegie (2012) at p. 10 (proliferators often use low-quality substitutes despite significant cost and performance penalties).
18 Wirtz (2010) at p. 282.
19 Albright (2010) at pp. 180, 187.
20 Carnegie (2012) at p. 9.
21 See, e.g., Hund and Seward (2008) at p. 15 (the Nuclear Suppliers Group and most of its member states do not share data with the International Atomic Energy Agency).
22 See Wirtz (2010) at pp. 274–9 (for the most part, industry exchanges with governments are still voluntary and consist mostly of informal meetings), 268–9 (recounting how unnamed European spy agency showed company that apparently innocuous order was part of a much larger and patently illicit project); Hund and Seward (2008) at p. 5; Carnegie (2012) at p. 9; see also Hund and Elkhamri (2005) at p. 5 (describing firms' reluctance to share information about suspicious inquiries or suspected front companies).
23 Joyner (2004) at p. 109.
24 Carnegie Fund (2012) at p. 9; cf. Wirtz (2010) at p. 255 (industry detected Iranian nuclear program and South African proliferators before regulators did).
25 Dupre (1992) at p. 46 ("catch-all" regulations require companies to seek licenses whenever they know or have reason to know that the inquiry has originated with an illicit end user); Joyner (2004) at p. 110 (catch-all legislation extends to "conscious disregard" or "willful avoidance of facts").
26 These notably include private (but government-funded) lists of known front companies. Wisconsin Project (n.d.(a)). The Wisconsin Project is supported by grants from the U.S. government and "several private foundations." Wisconsin Project (n.d.(b)).
27 Joyner (2004) at pp. 115–18 (courts have repeatedly held that regulations that fail to clearly specify criminal behavior are "void for vagueness" under the U.S. Constitution); Dupre at p. 46 (same).
28 Dupre (1992) passim.
29 Albright (2010) at pp. 88 and 98 (describing high failure rate of Libyan and Iranian centrifuges), 188 (proliferators deliberately accept low-performance substitutes to evade export rules); Langer (2013) at p. 6 (Iran operates its machines at 50 percent capacity to limit crashes); Lynch (2015) (Iran "settle[s] for cheaper made, less reliable substitutes by producers in Asia" and limited-performance, "low grade components" not found on nuclear control lists.).
30 Albright (2010) at pp. 46–9.

31 Wirtz (2010) at p. 263; Hund and Seward (2008) at p. 19 (quoting Berkeley Nucleonics' response to Leybold-style screening proposals: "Sounds interesting, but I don't have time to take it on; my company is too small").
32 Albright (2010) at pp. 180–1 (Leybold claims to have refused $50–60 million from 1990–2010); Hund and Seward (2008) at p. 1 (Leybold rejected orders worth €25 million from 1993–2003).
33 Hund and Seward (2008) at p. 14.
34 Id.
35 Carnegie (2012) at p. 9.
36 Wirtz (2010) at p. 275. Hund and Seward (2008) at p. 5.
37 Hund and Seward (2008) at p. 14 (describing how Leybold obtained detailed confidentiality and cybersecurity assurances from IAEA to facilitate sharing).
38 Hund and Elkhamri (2005) at p. 11 (arguing that a reformed system should allow industry to "green lane" exports to approved recipients).
39 Hund and Seward (2008) at pp. 1 (government can improve nuclear security by trading sterile "adversarial" and ineffective compliance activities for more industry effort toward creative solutions), 19 (quoting General Electric executive: firms are more likely to adopt proactive screening practices "if industry receives some benefit in return"), 1 (self-regulation creates potential for "enhanced trust [that] might result in a less adversarial approach"); Wirtz (2010) at p. 259.
40 Id. at p. 280.
41 See, generally, Albright (2010), pp. 47–56 et passim. Additional extended description of Iraqi and Iranian front companies can be found in Fitzpatrick (2007); Ottolenghi (2013); and ISIS (n.d.).
42 See Albright (2010 YES) at pp. 100–6 (describing small family machine shops that produced centrifuge components). Illicit budgets often support even larger enterprises. See, e.g., Albright (2010) at pp. 107 ($3.4 million contract for centrifuge parts accounted for 26 percent of Malaysian industrial group's annual revenue), 51 (Leybold's Hanau division received roughly one-fourth of its revenues from the Khan network in the late 1970s).
43 Id. at pp. 20, 80, 88.
44 Anon. (2010) (fewer than a dozen firms). Other estimates put the number at fewer than a half dozen. Anon. (2013).
45 Anon. (2010). While Iran started producing carbon fiber domestically in 2011, most analysts believe that the product lacks the quality needed to make rotors. Albright and Walrond (2011) at p. 3; Lynch (2015) (IAEA believes that "most" of the high-strength aluminum, carbon fiber, and managing steel used in Iran's centrifuges is "probably imported").
46 See, e.g., Sloan (2013) at Table 2 (quoting Composites Forecasts and Consulting LLC); Anon. (2008).
47 Readers can readily confirm this by comparing carbon-fiber manufacturers against press releases for individual Boeing and Airbus procurement programs.
48 Anon. (2010).
49 World Nuclear Assn. (2016).
50 The current incumbents are Toshiba, Hitachi, Mitsubishi (Japan), INVAP (Argentina), KEPCO (Korea), Areva (France), General Electric, Westinghouse (United States), and Rusatom (Russia). NuclearPrinciples.org (n.d.).
51 Wikipedia (n.d.(g)).

52 World Nuclear Assn. (2016).
53 Hund and Seward (2008) at p. 17.
54 Perkovich and Radzinsky (2012) at p. 18. Recent history, including the Fukushima accident and Germany's ensuing decision to denuclearize, shows that these fears have merit.
55 Id. at p. 7.
56 Id. at pp. 10, 20.
57 Id. at p. 14.
58 Id. at pp. 14, 15.
59 Id. at p. 15.
60 World Nuclear Assn. (2016).
61 Gosden (2016) (discussing Chinese state firm CGN's three-reactor deal in the UK).
62 Wikipedia (n.d. (b, c, f)).
63 Carnegie Fund (2012) at p. 9.
64 Maurer (2009).
65 Cesaroni, Gambardella, and Mariani (2007) at p. 38.
66 Spitz (2003) at p. 252.
67 Spitz (2003) at pp. 250, 257.
68 Id. at p. 328 ("Companies have gone in the direction of reducing – sometimes greatly – the number of vendors they deal with in order to facilitate procurement and exert greater pricing leverage"); Boccone (2003) at p. 89 97 ("The number of suppliers to any given customer is decreasing, with sole source supply contracts becoming more common. As the customer base consolidates and moves more toward sole source supply contracts, relationships selling at the local or plant level will become unnecessary and is already being replaced in some cases by master contracts at the corporate level"). For a discussion of Big Pharma's leverage over their chemical industry suppliers, see Challenger (2003) (reporting how large customers demand "almost annual price decreases" and free services from vendors) and Hume and Schmitt (2001) (documenting producers' claim that competition has cut margins from 15–20 percent to 10–15 percent).
69 Boccone (2003) at pp. 91 ("Al Schuman, the CEO of Ecolab, had it right when he said that the idea was to 'sell the account with our differentiated product and keep the account with our service'"), 94 (chemical industry executives believe that customer relations and services are far more important than any other factor including innovation, organizational flexibility, global reach, or operational excellence).
70 Boccone (2003) at pp. 86, 98–9 (describing examples in paper, autos, and healthcare). The fact that suppliers are able to maintain high margins suggests significant countervailing power. Id.
71 Maurer et al. (2009) at pp. 65–6; Nguyen (2005).
72 Boswell (2009).
73 See, e.g., Thomas.net (n.d.) (listing 62 bioreactor manufacturers).
74 Maurer et al. (2009).
75 Id. at p. 9.
76 Fischer and Maurer (2010).
77 Garfinkel et al. (2007); Maurer et al. (2009) at p. 25.
78 Ferguson and Lubenau (2003); World Nuclear Assn. (2016).
79 Hund and Seward (2008) at p. A-2.
80 Ferguson and Lubenau (2003) (1998 estimate).

81 Id.
82 Hund and Seward (2008) at p. 11.
83 Microsoft (2007).
84 Higgins (2011); Kan (2011).
85 Wikipedia (n.d. (k)). The Chinese government recruits many of its "cybermilitias" from IT workers at banks and similar venues that are almost entirely insulated from Western business. Carr (2011) at p. 632.
86 RussSoft Association (2015) at p. 42 (slightly less than two-thirds of all Russian software companies participate in the export market).
87 Bennett (2015) (hackers work for Russian state in exchange for official tolerance, including a willingness to drop criminal charges).
88 Anon. (2013); Hvistendahl (2010).
89 Forsythe and Sanger (2015).
90 Maurer, Lucas, and Terrell (2007) at pp. 18–19.
91 Id.; Maurer (2011a) at p. 1414–15. Our nuclear physics (Ch. 1.2.1) and Asilomar (Ch. 5.1) examples document multiple instances where scientists deferred to colleagues despite their own personal belief that their experiments should go forward.
92 See e.g., Maurer et al. (2011) (NIH, SERCEB, UC Berkeley advice portals); Simirenko et al. (2015) (Department of Energy portal).
93 See, e.g., Maurer, Lucas, and Terrell (2007) at pp. 21–2.
94 See, e.g. Maurer, Firestone and Scriver (2000).

References

4C Association. 2007. *Annual Report*, www.4c-coffeeassociation.org/uploads/media/4C_Annual_Report_2007.pdf.

4C Association. 2010. *Annual Report*, www.4c-coffeeassociation.org/uploads/media/4C_Annual Report2009_web_lr_en.pdf.

Abbott, KW, JF Green, and RO Keohane. 2016. "Organizational Ecology and Organizational Diversity in Global Governance," *International Organization/First View Article* (Apr.): 1–31.

Abbott, KW and D Snidal. 1996. "International 'Standards' and International Governance," *Journal of European Public Policy* 8(3): 345–370.

2009a. "Strengthening International Regulation Through Transnational New Governance: Overcoming the Orchestration Deficit," *Vanderbilt Journal of Transnational Law* 42: 501.

2009b. "The Governance Triangle: Regulatory Standards Institutions and the Shadow of the State," pp. 44–88 in W Mattli and N Woods (eds.), *The Politics of Global Regulation* (Princeton University Press: Princeton NJ).

Abreu, D, D Pearce, and E Stacchetti. 1990. "Toward a Theory of Discounted Repeated Games with Imperfect Monitoring," *Econometrica* 58(5): 1041–1063.

Administrative Conference of the United States (ACUS). 1978. "Recommendation 78-4, Federal Agency Interaction with Private Standard-Setting Organizations in Health and Safety Regulation."

Akerlof, GA. 1970. "The Market for 'Lemons': Quality Uncertainty and the Market Mechanism," *Quarterly Journal of Economics* 84(3): 488–500.

Albright, D. 2010. *Peddling Peril* (Free Press: New York).

Albright, D and M Hibbs. 1992. "Iraq's Bomb: Blueprints and Artifacts," *Bulletin of the Atomic Scientists* (Jan–Feb): 30.

Albright, D and C Walrond. 2011. *Iran's Advanced Centrifuges* (ISIS: Washington DC).

Aldhous, P. 2005. "The Bioweapon Is in the Post," *New Scientist* (Nov. 9).

2006. "Synthetic Biologists Reject Controversial Guidelines," *New Scientist* (May 23), www.newscientist.com/article/dn9211-synthetic-biologists-rejectcontroversial-guidelines.html.

Anchustegui, IH. 2015. "Competition Law through an Ordoliberal Lens," *Oslo Law Review* 2(3): 139.

Anderton, Alain. 2000. *Economics*, 3rd ed. London: Pearson Education Ltd.

Anon. 1963. "Developments in the Law: Judicial Control of Actions of Private Associations: Association Action Affecting Nonmembers," *Harvard Law Review* 76: 1037.

2004. "The First International Meeting on Synthetic Biology, Synthetic Biology 1.0, June 10–12, 2004," openwetware.org/images/7/79/SB1.0_overview.pdf.

2005a. "Editorial: Genes for Sale," *New Scientist* (Nov. 12): 5, www.precaution.org/lib/05/genes_for_sale.051112.htm.

2005b. "Craig Venter Institute Press Release: Major New Policy Study Will Explore Risks, Benefits of Synthetic Genomics," *Biospace* (Jun. 28), www.biospace.com/news_story.aspx?NewsEntityId=20467320.

2008a. "Carbon Fibre: Sustained Increase in Production Capacities," *JEC Magazine* 41 (Jun.), www.jeccomposites.com/knowledge/international-composites-news/carbon-fibre-sustained-increase-production-capacities.

2008b. Editorial, "Pathways to Security," *Nature* 455: 432.

2008c. "A Stitch in Time: How Companies Manage Risks to Their Reputation," *The Economist* (Jan. 17).

2010a. "Global and China Carbon Fiber Industry Report 2009–2010," www.researchinchina.com/Htmls/Report/2010/5939.html.

2010b. "Has Emmer Donation Cost Target on Stock Market?" *Pioneer Press* (Aug. 3), www.twincities.com/2010/08/03/has-emmer-donation-cost-target-on-stock-market.

2011. "Scientists Asked Not to Reveal Bird Flu Details," *Wall Street Journal* (Dec. 22).

2013a. "Market Outlook: Surplus in Carbon Fiber's Future?" *CompositesWorld* (Mar. 2013), www.compositesworld.com/articles/market-outlook-surplus-in-carbon-fibers-future.

2013b. "Masters of the Cyber-Universe," *The Economist* (Apr. 6).

Answers.com n.d. "How Wide Is a Human Hair?" wiki.answers.com/Q/What_is_the_average_thickness_of_a_human_hair.

AstraZeneca. 2009. "Code of Conduct," www.astrazeneca.com/Responsibility/Code-policies-standards/Code-of-Conduct.

Auld, G. 2010. "Assessing Certification as Governance: Effects and Broader Consequences for Coffee," *Journal of Environment and Development* 19: 216.

Auld, G, S Bernstein, and B Cashore. 2008. "The New Corporate Social Responsibility," *Annual Review of Environment and Resources* 33: 413–435.

Auld, G, L Bozzi, B Cashore, K Levin, and S Renckens. 2007. "Can Non-State Governance 'Ratchet Up' Global Standards? Assessing Their Indirect Effects and Evolutionary Potential," *Review of European Community and International Environmental Law* 16(2): 158.

Auld, G, B Cashore, C Balboa, L Bozzi, and S Renckens. 2010. "Can Technological Innovations Improve Private Regulation in the Global Economy?" *Business and Politics* 12(3).

Auld, G, B Cashore, and D Newsom. 2002. "Perspectives on Forest Certification: A Survey Examining Differences Among the US Forest Sectors' Views of Their Forest Certification Alternatives," pp. 271ff in LD Teeter, B Cashore, and D Zhang (eds.), *Forest Policy for Private Forestry* (CABI Publishing: Oxford and New York).

Bäckstrand, K. 2006. "Democratizing Global Environmental Governance? Stakeholder Democracy after the World Summit on Sustainable Development," *European Journal of International Relations* 12(4): 467–498.

Balleisen, EJ and M Eisner. 2009. "The Promise and Pitfalls of Co-Regulation: How Governments Can Draw on Private Governance for Public Purpose" pp. 127–149 in D Moss and J Cisternino (eds.), *New Perspectives on Regulation* (The Tobin Project: Cambridge MA).

Balmer, A and P Martin. 2008. "Synthetic Biology: Social and Ethical Challenges" (Institute for Science and Society), www.synbiosafe.eu/uploads///pdf/synthetic_biology_social_ethical_challenges.pdf.

Balter, M. 1999. "EMBL Faces Huge Bill Following Adverse Pay Dispute Ruling," *Science* 286: 1058–1059.

2001."Windfall for European Data Bank," *Science* 292: 1275.

Banjo, S. 2014. "Inside Nike's Struggle to Balance Cost and Worker Safety in Bangladesh," *Wall Street Journal* (Apr. 21).

Barinaga, Marcia. 2003. "Asilomar Revisited: Lessons for Today?," *Science* 287: 5458.

Baron, DP. 2003. "Private Politics," *Journal of Economics and Management Strategy* 12: 31.

Bartley, T. 2011. "Transnational Governance as the Layering of Rules: Intersections of Public and Private Standards," *Theoretical Inquiries in Law* 12(2).

BASF. n.d. "Guide to Safe Manufacture and for Activities Involving Nanoparticles at Workplaces in BASF AG," www.basf.com/group/corporate/en/function/conversions:/publish/content/sustainability/dialogue/in-dialogue-with-politics/nanotechnology/images/BASF_Guide_to_safe_manufacture_and_for_activities_involving_nanoparticles.pdf.

Beisheim, M and C Kaan. 2010. "Transnational Standard-Setting Partnerships in the Field of Social Rights: The Interplay of Legitimacy, Institutional Designs, and Process Management," pp. 122 ff in M Bexell and U Mörth (eds.), *Democracy and Public-Private Partnerships in Global Governance* (Palgrave: Basingstoke UK).

Bennett, C. 2015. "Kremlin's Ties to Russian Cyber Gangs Sow US Concerns," *The Hill* (Oct. 11), thehill.com/policy/cybersecurity/256573-kremlins-ties-russian-cyber-gangs-sow-us-concerns.

Berg, P. 2008. "Meetings that Changed the World: Asilomar 1975," *Nature* 455: 290–291.

Berg, P, D Baltimore, HW Boyer, SN Cohen, RW Davis, DS Hogness, D Nathans, R Roblin, JD Watson, S Weissman, and ND Zinder. 1974. "Potential Biohazards of Recombinant DNA Molecules," *Science* 185: 303.

Berg, P and MF Singer. 1995. "The Recombinant DNA Controversy: Twenty Years Later," *PNAS* 92: 9011–9013.

Berg, S and R Cantrell. 1999. "Sustainable Forestry Initiative: Toward a Higher Standard," *Journal of Forestry* 97: 33.

Berger, K. 2012. *Bridging Science and Security for Biological Research: A Dialogue between Universities and the Federal Bureau of Investigation. Report of a Meeting February 21–22, 2012*, www.aaas.org/sites/default/files/AAAS-AAU-APLU-FBI-Final-Report_Feb.pdf.

2013. *Meeting Report: Bridging Science and Security for Biological Research: Personnel Security Programs* (National Assn. for the Advancement of Science: Washington).

Bernauer, H, J Christopher, W Deininger, M Fischer, P. Habermeier, K Heumann, S Maurer, H Schwer, P Stähler, and T Wagner. 2008. *Report on the Workshop "Technical Solutions for Biosecurity in Synthetic Biology"* (International Association Synthetic Biology: mimeo), www.ia-sb.eu/go/synthetic-biology/activities/press-area/press-information/iasb-report-on-biosecurity-and-biosafety.

Berners-Lee, T and M Fischetti. 2000. *Weaving the Web: The Original Design and Ultimate Destiny of the World Wide Web* (Harper Business: New York).
Bernstein, S. 2011. "Legitimacy in Intergovernmental and Non-State Global Governance," *Review of International Political Economy* 18(1): 17–51.
Bernstein, S and B Cashore. 2007. "Can Non-State Global Governance Be Legitimate? An Analytical Framework," *Regulation & Governance* 1: 347–371.
Berry, JM. 1989. "Subgovernments, Issue Networks, and Political Conflict," pp. 239–260 in RA Harris and S Miklis (eds.), *Remaking American Politics* (Westview Press: Boulder CO).
Besley, T and M Ghatak. 2007. "Retailing Public Goods: The Economics of Corporate Social Responsibility," *Journal of Public Economics* 91: 1645.
Biospace.com. 2010. "GENEART AG Will Become Part of the US-Based Biotechnology Company Life Technologies" (Apr. 9), www.biospace.com/News/geneart-ag-will-become-part-of-the-us-based/176354.
Bisson, TA. 1954. *Zaibatsu Dissolution in Japan* (University of California Press: Berkeley).
Black, J. 2008. "Constructing and Contesting Legitimacy and Accountability in Polycentric Regulatory Regimes," *Regulation & Governance* 2: 137–164.
Boccone, A. 2003. "Specialty Chemicals," pp. 85–110 in PH Spitz (ed.), *The Chemical Industry at the Millennium: Maturity, Restructuring, and Globalization* (Chemical Heritage Foundation: Philadelphia PA).
Bodansky, DM. 2008. "Legitimacy in International Environmental Law," pp. 704–726 in D Bodansky, J Brunnée, and E Hey (eds.), *Oxford Handbook of International Environmental Law* (Oxford University Press: New York).
Borelli, D. 2016. "The Financial Consequences of Target's New Bathroom Policy," *Conservative Review* (May 4), www.conservativereview.com/commentary/2016/05/financial-consequences-of-targets-new-bathroom-policy.
Bornmann, L. 2011. "Scientific Peer Review," *Annual Review of Information Science and Technology* 45(1): 197–245.
Börzel, TA, CR Thauer, and J Hönke. 2011. "Racing to the Top? Regulatory Competition Among Firms in Areas of Limited Statehood," pp. 144ff in T Risse (ed.), *Governance Without A State* (Columbia University Press: New York).
Boswell, C. 2009. "Microreactors Gain Popularity Among Producers," *ICIS Chemical Business*, www.icis.com/resources/news/2009/05/04/9211877/microreactors-gain-popularity-among-producers.
Bowen, D. 2012. "Cost of Government Day: 2012 Report" costofgovernment.org/files/files/COGD2012_hi%20res.pdf.
Bowman, DM and GA Hodge. 2009. "Counting on Codes: An Examination of Transnational Codes as a Regulatory Governance Mechanism for Nanotechnologies," *Regulation & Governance* 3: 145–164.
Breton, A and R Wintrobe. 1982. *The Logic of Bureaucratic Conduct: An Economic Analysis of Competition, Exchange, and Efficiency in Private and Public Organizations* (Cambridge University Press: New York).
Brokenpipeline.org. 2008. "A Broken Pipeline? Flat Funding of the NIH Puts a Generation of Science at Risk," www.brokenpipeline.org.
Brousseau, E and E Raynaud. 2011. "Climbing the Hierarchical Ladders of Rules: A Life-Cycle Theory of Institutional Evolution," *Journal of Economic Behavior and Organization* 79: 65.

Brown, R and Zhang, D. 2005. "The Sustainable Forestry Initiative's Impact on Stumpage Markets in the US South," *Canadian Journal of Forest Resources* 35: 2056.

Buchanan, JM and G Tullock. 1974 [1962]. *The Calculus of Consent* (University of Michigan Press: Ann Arbor MI).

Bügl, H, JP Danner, RJ Molinari, JT Mulligan, H-O Park, B Reichert, et al. 2007. "DNA Synthesis and Biological Security," *Nature Biotechnology* 25: 627–629.

Bumas, Lester O. 1999. *Intermediate Microeconomics: Neoclassical and Factually-Oriented Models*. New York: M.E. Sharpe, Inc.

Burton, TE. 1906. *John Sherman* (Houghton Mifflin: New York).

Bush, SR, H Toonen, P Oosterveer, and APJ Mol. 2012. "The 'Devils Triangle' of MSC Certification: Balancing Credibility, Accessibility and Continuous Improvement," *Marine Policy* 37: 288.

Bush, V. 1945. *Science, the Endless Frontier* (US Government Printing Office: Washington DC).

Büthe, T. 2010a. "Engineering Uncontestedness? The Origins and Institutional Development of the International Electrotechnical Commission (IEC)," *Business and Politics* 12(3).

2010b. "Global Private Politics: A Research Agenda," *Business and Policy* (Oct.).

2010c. "Private Regulation in the Global Economy: A (P)Review," *Business and Policy* 12(3): 1–38.

2010d. "Private Regulation in the Global Economy: Guest Editor's Note," *Business and Policy* 12(3): 1–2.

Büthe, T and W Mattli. 2011. *The New Global Rulers: The Privatization of Regulation in the World Economy* (Princeton University Press: Princeton NJ).

Butler, D. 1999. "Life Science Facilities in Crisis as Brussels Switches Off Funding," *Nature* 402: 3–4.

Cabral, LMB. 2005. *The Economics of Trust and Reputation: A Primer* (Preliminary), pages.stern.nyu.edu/~lcabral/reputation/Reputation_June05.pdf.

Cafaggi, F and A Renda. 2012. "Public and Private Regulation: Mapping the Labyrinth," CEPS Working Paper No. 370.

California Environmental Associates. 2012. "A Literature and State-of-Knowledge Review of Fisheries Certification and Standards," pp. A-70–A-87 in Steering Committee of the State of Knowledge Assessment of Standards and Certification, *Toward Sustainability: The Roles and Limitations of Certification*.

Calliess, GP and M Renner. 2009. "Transnationalizing Private Law," *The German Law Journal* 10: 1341–1356.

Campbell, E, B Clarridge, M Gokhale, L Birenbaum, S Hilgartner, N Holtzman, and D Blumenthal. 2002. "Data Withholding in Academic Genetics: Evidence from a National Survey," *Journal of the American Medical Association* 287: 473.

Campbell, H and R Le Heron. 2007. "Supermarkets, Producers, and Audit Technologies: The Constitutive Micro-Politics of Food, Legitimacy, and Governance," pp. 131–153 in D Burch and G Lawrence (eds.), *Supermarkets and Agri-Food Supply Chains* (Elgar: Cheltenham UK).

Camus, A. 1955 [1942]. *The Myth of Sisyphus* (Hamish Hamilton: London).

Carlson, R. 2005. "Synthetic Biology 1.0," *Futurebrief*, www.futurebrief.com/robertcarlsonbio.pdf.

Carman, T. 2012. "WTO: 'Dolphin-Safe' Label Discriminates Against Mexico," *Washington Post* "All We Can Eat" Blog (May 16), www.washingtonpost.com/blogs/all-we-can-eat/post/wto-dolphin-safe-label-discriminates-against-mexico/2012/05/16/gIQAwnCcUU_blog.html.

Carnegie Fund for Peace. 2012. *Guiding Principles and Nuclear Proliferation: An Industry-Based Approach to Strengthening Nonproliferation*, library.fundforpeace.org/library/ttcvr1204-threatconvergence-guidingprinciples-02d.pdf.

Carr, J. 2011. *Inside Cyberwarfare* (2d ed.) (O'Reilly: Sebastopol CA).

Carter, GM, WD Cooper, CS Lai, and DMS March. 1978. *The Consequences of Unfunded NIH Applications for the Investigator and His Research*, RAND Report No. R-2229-NIH (RAND: Santa Monica CA).

Cashore, B. 2004. "Environmental Governance: How Non-State Market-Driven (NSMD) Governance Systems Gain Rule-Making Authority," *Governance* 15(4): 503–529.

Cashore, B and G Auld. 2012. "Forestry Review," pp. A-88–A-124 in Steering Committee of the State of Knowledge Assessment of Standards and Certification, *Toward Sustainability: The Roles and Limitations of Certification*.

Cashore, B, G Auld, and D Newsom. 2004. *Governing Through Markets: Forest Certification and the Emergence of Non-State Authority* (Yale University Press: New Haven CT).

Cater, D. 1964. *Power in Washington* (Random House: New York).

Cello, J, AV Paul, and E Wimmer. 2002. "Chemical Synthesis of Polio-Virus cDNA: Generation of Infectious Virus in the Absence of a Natural Template," *Science* 297: 1016–1018.

Cesaroni, B, A Gambardella, and M Mariani. 2007. "The Evolution of Networks in the Chemical Industry, pp. 21–52 in L Galamos, T Hikino, and V Zamagni, *The Global Chemical Industry in the Age of the Petrochemical Revolution* (Cambridge University Press: New York).

Chait, J. 2012. "The Vast Left-Wing Conspiracy Is on Your Screen," *New York Magazine* (Aug. 19).

Challenger, C. 2003. "The Balancing Act of Small to Medium Sized Custom Manufacturers," *Chemical Market Reporter* (Apr. 14).

Check Hayden, E. 2003. "US Officials Urge Biologists to Vet Publications for Bioterror Risk," *Nature* 421: 197.

— 2009a. "Gene-Makers Form Security Coalition," *Nature News* (Nov. 18), www.nature.com/news/2009/091118/full/news.2009.1095.

— 2009b. "Keeping Genes Out of Terrorists' Hands," *Nature*, 461: 22.

Chilvers, J and P Macnaghten. 2011. "The Future of Science Governance: A Review of Public Concerns, Governance, and Institutional Response," www.sciencewise-erc.org.uk/cms/assets/Uploads/Project-files/Future-of-Science-Governance-Lit-Review-April-new.pdf.

Chubin, DE and EJ Hackett. 1990. *Peerless Science: Peer Review and US Science Policy* (SUNY Press: Albany NY).

Church, G. 2005. "Let Us Go Forth and Safely Multiply," *Nature* 438: 423.

Committee on Advances in Technology and the Prevention of Their Application to Next Generation Biowarfare Threats. 2006. *Globalization, Biosecurity and the Future of the Life Sciences* (National Research Council: Washington DC).

Committee on Research Standards & Practices to Prevent the Destructive Application of Biotechnology. *See* Fink Committee

Committee on Scientific Milestones for the Development of a Gene Sequence-Based Classification System for the Oversight of Select Agents. 2010. *Sequence-Based Classification of Select Agents: A Brighter Line* (National Research Council: Washington DC).

Conroy, ME. 2007. *Branded! How the "Certification Revolution" Is Transforming Global Corporations* (New Society Publishers: Gabriola Island Canada).

Cummins, A. 2004. "The Marine Stewardship Council: A Multi-Stakeholder Approach to Sustainable Fishing," *Corporate Social Responsibility and Environmental Management* 11: 85.

Dahl, RA. 1957. "The Concept of Power," *Behavioral Science* 2: 201.

Dando, M. 2010. "Synthetic Biology: Harbinger of an Uncertain Future?" *Bulletin of the Atomic Scientists* (Aug. 16), www.thebulletin.org/web-edition/columnists/malcolm-dando/synthetic-biology-harbinger-of-uncertain-future.

Davies, P and D Lynch. 2002. *Routledge Companion to Fascism and the Far Right* (Routledge: London).

Dawson, NL and K Segerson. 2008. "Voluntary Agreement with Industries: Participation Incentives with Industry-Wide Targets," *Land and Economics* 84: 97.

de Vriend, H. 2006. "Constructing Life: Early Social Reflections on the Emerging Field of Synthetic Biology," Rathenau Institute Working Doc. No. 97 (Rathenau Inst.: The Hague), depot.knaw.nl/4935/1/WED97_Constructing_Life_2006.pdf.

DeLoeker, J and J Eeckhout. 2017. "The Rise of Market Power and Macroeconomic Implications," National Bureau of Economic Research Working Paper No. 23687 (NBER: Boston, MA). Available at http://www.nber.org/papers/w23687.

Dingwerth, K. 2007. *The New Transnationalism: Transnational Governance and Democratic Legitimacy* (Palgrave: Houndmills).

Dixon, RG Jr. 1978. *Standards Development in the Private Sector: Thoughts on Interest Representation and Procedural Fairness* (National Fire Protection Assn.: Quincy MA).

Djama, M, E Fouilleux, and I Vagneron. 2011. "Standard-Setting, Certifying and Benchmarking: A Governmentality Approach to Sustainability Standards in the Agro-Food Sector," ch. 8 in S Ponte, P Gibbon, and J Vestergaard (eds.), *Governing through Standards: Origins, Drivers and Limitations* (Palgrave: New York).

Downs, A. 1957. *An Economic Theory of Democracy* (Harper: New York).

Doyle, C. 1997. "Self Regulation and Statutory Regulation," *Business Strategy Review* 8: 35.

DuPont Chemical Co. and Environmental Defense Fund. 2007. "NanoRisk Framework," www.edf.org/documents/6496_Nano%20Risk%20Framework.pdf.

Dupre, AP. 1992. "Foreign Duty: Export Control Goes Private," *Business Law Today* 2: 46.

Dworkin, R. 1987. *Law's Empire* (Harvard University Press: Cambridge MA).

Earth Island Institute. 2011. *International Tuna Monitoring Program 2011 Annual Report*, www.earthisland.org/dolphinSafeTuna/assets/2011MonitorReport.pdf.

n.d.(a). "Approved Dolphin Safe Tuna Processing Companies and Fishing Companies," www.earthisland.org/dolphinSafeTuna/DolphinSafe Canners.html.

n.d.(b). "International 'Dolphin Safe' Standards For Tuna," www.earthisland.org/immp/Dol_Safe_Standard.html.

n.d.(c). Web Page: "International Dolphin Safe Monitoring Program," www.earthisland.org/dolphinSafeTuna.

Eberlein, B, KW Abbott, J Black, E Meidinger, and S Wood. 2014. "Transnational Business Governance Interactions: Conceptualization and Framework for Analysis," *Regulation & Governance* (2014) 8: 1–21.

Edmunds, T and R Wheeler. 2009. "Setting Priorities: Assessing Threats and Identifying Cost-Effective Responses to WMD Terrorism," pp. 191–210 in SM Maurer (ed.), *WMD Terrorism: Science and Policy Choices* (MIT Press: Cambridge MA).

Eisenstein, M. 2010. "Synthetic DNA Firms Embrace Hazardous Agents Guidance But Remain Wary of Automated 'Best Match,'" *Nature Biotechnology* 28: 1225.

El Baradei, M. 2006. "Rethinking Nuclear Safeguards," *Washington Post* (Jun. 14).

Elhauge, ER. 1991. "The Scope of Antitrust Process," *Harvard Law Review* 104: 667.

Enloe, C. 2017. " Liberal Boycott of Sean Hannity Backfires and Puts Multiple MSNBC Stars in Crosshairs, " *The Blaze* (May 29, 2017) available at www.theblaze.com/news/2017/05/29/liberal-boycott-of-sean-hannity-backfires-and-puts-multiple-msnbc-stars-in-crosshairs/.

ETC Group. 2006. "Backgrounder: Open Letter on Synthetic Biology" (May 17), www.etcgroup.org/en/node/11.

European Cetacean ByCatch Campaign. n.d. Web Page: "The Tuna Boycott That Led to the Dolphin Safe Label," www.eurocbc.org/page322.htmls.

FAO. *See* United Nations Food & Agricultural Organisation

Farrell, J and G Saloner. 1988. "Coordination Through Committees and Markets," *RAND Journal of Economics* 19: 235.

Ferguson, CD and JO Lubenau. 2003. "Securing US Radioactive Sources," *Issues in Science and Technology Policy* XX(1), issues.org/20–1/ferguson.

Feynman, M. 2005 [1966]. *Perfectly Reasonable Deviations (from the Beaten Track): The Letters of Richard P. Feynman* (Basic Books: New York).

Fink Committee: *See* National Academy of Sciences Committee on Genomic Databases for Bioterrorism

Fiorino, DJ. 2010. *Voluntary Initiatives, Regulation and Nanotechnology Oversight: Charting a Path*, www.nanotechproject.org/process/assets/files/8347/pen-19.pdf.

Fischer, C and TP Lyon. 2014. "Competing Environmental Labels," *Journal of Economics and Management Strategy* 23: 692.

Fischer, M and SM Maurer. 2010. "Harmonizing Biosecurity Oversight for Gene Synthesis," *Nature Biotechnology* 28: 20–22.

Fischetti, M. 2009. "The Mind Behind the Web." *Scientific American* (Mar. 12).

Fitzpatrick, M. 2007. *Nuclear Black Markets: Pakistan, AQ Khan, and the Rise of Proliferation Networks* (International Institute for Strategic Studies: Washington DC).

Foreman, CH. 1988. *Signals from the Hill: Congressional Oversight and the Challenge of Social Regulation* (Yale University Press: New Haven CT).

ForestEthics. 2010. *SFI: Certified Greenwash – Inside the Sustainable Foresty Initiative's Deceptive Eco-Label*, forestethics.org//sites/forestethics.huang.radicaldesigns.org/files/SFI-Certified-Greenwash-Report-ForestEthics.pdf.

Forest Stewardship Council. 2012. "FSC International Standard: FSC Principles and Criteria for Forest Stewardship 7," ic.fsc.org/download.fsc-std-01-001-v5-0-revised-principles-and-criteria-for-forest-stewardship.a-1780.pdf.

   n.d. Web Page: "FSC General Assembly," www.ga.fsc.org/generalassembly.html.

Forsythe, M and DE Sanger. 2015. "China Calls Hacking of U.S. Workers' Data a Crime, Not a State Act" (Dec. 2).

Frederickson, DS. 1991. "Asilomar and Recombinant DNA: The End of the Beginning," pp. 258–307 in KE Hanna (ed.), *Biomedical Politics* (National Academy Press: Washington DC).

Friedrichs, S. 2007. "Deciphering Nanotechnology Codes," *Nanotechnology Now*, www.nanotech-now.com/columns/?article=093.

FSC. *See* Forest Stewardship Council.

FTC. *See* U.S. Federal Trade Commission.

Fuchs, D and A Kalfagianni. 2010. "The Causes and Consequences of Private Food Governance," *Business and Policy* 12(3): 5.

Fuchs, D, A Kalfagianni, and M Arendtsen. 2009. "Retail Power, Private Standards, and Sustainability in the Global Food System," pp. 29 ff. in J Clapp and D Fuchs (eds.), *Corporate Power in Global Agrifood Governance* (MIT Press: Cambridge MA).

Fuchs, D, A Kalfagianni, and T Havinga. 2008. "Actors in Private Food Governance: The Legitimacy of Retail Standards and Multistakeholder Initiatives with Civil Society Participation," *Agriculture and Human Values* 28: 353.

Fudenberg, D and E Maskin. 1986. "The Folk Theorem in Repeated Games With Discounting or Incomplete Information," *Econometrica* 54(3): 533–554.

Furger, F. 1997. *Accountability and Systems of Self-Governance: The Case of the Maritime Industry, Law and Policy* 19: 445.

Garfinkel, M, D Endy, G Epstein, and R Friedman. 2007. *Synthetic Genomics: Options for Governance* (Center for Security and International Studies: Washington DC), www.synbiosafe.eu/uploads/pdf/Synthetic%20Genomics%20Options%20for%20Governance.pdf.

Garvin, DA. 1983. "Can Self-Regulation Work?" *California Management Review* 25: 37.

Geiger, RL. 1986. *To Advance Knowledge: The Growth of American Research Universities, 1900–1940* (Oxford University Press: New York).

Gerber, DJ. 1994. "Constitutionalizing the Economy: German Neo-liberalism, Competition Law and the 'New' Europe," *American Journal of Comparative Law* 42: 25.

Gereffi, G. 1994. "The Organization of Buyer-Driven Global Commodity Chains: How U.S. Retailers Shape Overseas Production Networks," in pp. 95–122 G Gereffi and M Korzeniewicz (eds.), *Commodity Chains and Global Capitalism*, (Praeger: Westport CT).

1999. "International Trade and Industrial Upgrading in the Apparel Commodity Chain," *Journal of International Economics* 48(1): 37–70.

Gereffi, G, J Humphrey, and T Sturgeon. 2005. "The Governance of Global Value Chains," *Review of International Political Economy* 12(1): 78–104.

GFSI. *See* Global Food Safety Initiative.

Gilmore, J. 2008. "Case Study 3: MSC Certification of the Alaska Pollock Fishery," pp. 269–284 in T Ward and B Philips (eds.), *Seafood Ecolabelling: Principles and Practice* (Wiley: Hoboken NJ).

Global Food Safety Initiative. 2013. *GFSI Benchmarking Process*, www.mygfsi.com/gfsi-benchmarking-general/applications-update.html.

2016. "GFSI Guidance Document Sixth Edition Overview," www.mygfsi.com/technical-resources/guidance-document.html.

n.d. "What Is GFSI?" www.mygfsi.com/about-gfsi.html.

GLOBALG.A.P. n.d.(a). "Technical Committees," www.globalgap.org/uk_en/who-we-are/governance/technical-committees.

n.d.(b). "The GLOBALG.A.P. Database," www.globalgap.org/uk_en/what-we-do/the-gg-system/GLOBALG.A.P.-Database.

Golan, E, F Kuchler, and L Mitchell. 2001. "Economics of Food Labeling," *Journal of Consumer Policy* 24: 117.

Gollan, J. 2011. "Lab Fight Raises U.S. Security Issues," *New York Times* (Oct. 22).
Gosden, E. 2016. "China to Build Nuclear Reactor in Essex After Hinkley Deal Approved," *The (UK) Telegraph* (Sept. 15).
Grant, RW and RO Keohane. 2005. "Accountability and Abuses of Power in World Politics," *American Political Science Review* 99(1): 29–43.
Green, EJ and RH Porter. 1984. "Noncooperative Collusion Under Imperfect Price Information," *Econometrica* 52(1): 87–100.
Greenberg, A. 2015. "Here's a Spy Firm's Price List for Secret Hacker Techniques," *Wired* (Nov. 18), www.wired.com/2015/11/heres-a-spy-firms-price-list-for-secret-hacker-techniques.
Grushkin, D. 2010. "Synthetic Bio, Meet 'FBIo,'" *The Scientist* 24: 44.
Guevara, E. 1961. *Guerilla Warfare* (University of Nebraska Press: Lincoln NE).
Gulbrandsen, LH 2009. "The Emergence and Effectiveness of the Marine Stewardship Council," *Marine Policy* 33: 654–660.
    2014. "Dynamic Governance Interactions: Evolutionary Effects of State Responses to Non-State Certification Programs," *Regulation & Governance* 8: 74–92.
Gunningham, N and J Rees. 1997. "Industry Self-Regulation," *Law & Policy* 19: 373.
Gupta, AK and LJ Lad. 1983. "Industry Self-Regulation: An Economic, Organizational, and Political Analysis," *Academy of Management Review* 8(3): 416–425.
Guston, DH. 2003. "The Expanding Rule of Peer Review Processes in the United States," pp. 81–97 in P Shapira and S Kuhlmann (eds.), *Learning from Science and Technology Policy Evaluations: Experiences in the United States and Europe* (Elgar: Cheltenham UK).
Hale, T. 2011. "Marine Stewardship Council," pp. 308ff in T Hale and D Held (eds.), *Handbook of Transnational Governance: Institutions and Innovations* (Wiley: Hoboken NJ).
Hamilton, RW. 1982. "Prospects for the Nongovernmental Development of Regulatory Standards," *American University Law Review* 32(2): 455.
Harris, R and J Paxman. 2002. *A Higher Form of Killing: The Secret History of Chemical and Biological Warfare* (Random House: New York).
Harris, S and R Germaine. 2002. "Improving Forest Management Through the Supply Chain: An Assessment of Wood Procurement Management Systems in the Forest Products Industry," pp. 259–269 in LD Teeter, B Cashore, and D Zhang (eds.), *Forest Policy for Private Forestry* (CABI: Wallingford UK).
Havinga, T. 2006. "Private Regulation of Food Safety by Supermarkets," *Law and Policy* 28: 515.
    2008. "Actors in Private Food Regulation: Taking Responsibility or Passing the Buck to Someone Else?" Nimigen Sociology of Law Working Paper Series No. 2008/1.
Health and Human Services, U.S. Department of *See* HHS.
Heilbron, JL and RW Seidel. 1990. *Lawrence and His Laboratory: A History of the Lawrence Berkeley Laboratory*, vol. 1 (University of California: Berkeley).
Hellman, A, MN Oxman, and R Pollack. 1973. *Biohazards in Biological Research: Proceedings of a Conference at Asilomar, January 22–24, 1973* (Cold Spring Harbor Laboratory: Cold Spring Harbor NY).
Héritier, A and M Rhodes (eds.). 2011. *New Modes of Governance in Europe: Governing in the Shadow of Hierarchy* (Palgrave: New York).
HHS. See U.S. Dept. of Health and Human Services.

Hicks, JR. 1935. "Annual Survey of Economic Theory: The Theory of Monopoly." *Econometrica* 3(1).

Higgins, KJ. 2011. "A Call to Disarm Black Hat Hackers in China," *Dark Reading*, www.darkreading.com/vulnerabilities---threats/a-call-to-disarm-black-hat-hackers-in-china/d/d-id/1136365.

Hobson, C. 2009. "Beyond the End of History: The Need for a 'Radical Historicisation' of Democracy in International Relations," *Millennium* 37(3): 631–657.

Hofstadter, R. 1991 [1964]. "What Happened to the Antitrust Movement?" pp. 20–31 in ET Sullivan (ed.), *The Political Economy of the Sherman Act: The First One Hundred Years* (Oxford University Press: New York).

Holmes, OW. 1991 [1881] *The Common Law* (Dover: New York).

Home Depot. 2010. Press Release: "The Home Depot Introduces Eco Options" (Apr. 17), www.newswire.ca/en/story/84479/the-home-depot-introduces-eco-options.

   n.d.(a) Web Page: "Frequently Asked Questions," corporate.homedepot.com/CorporateResponsibility/Environment/WoodPurchasing/Pages/FAQs.aspx.

   n.d.(b). Web Page: "Wood Purchasing Policy," corporate.homedepot.com/CorporateResponsibility/Environment/WoodPurchasing/Pages/default.aspx.

Hotelling, H. 1929. "Stability in Competition," *Economic Journal*, 39(153): 41–57.

Howes, R. 2008. "The Marine Stewardship Council Programme," pp. 81–103 in T Ward and B Philips (eds.), *Seafood Ecolabelling: Principles and Practice* (Wiley: Hoboken NJ).

Hume, C and W Schmitt. 2001. "Pharma's Prescription," *Chemical Week* (Apr. 11): 21.

Humphreys, D. 1996. *Forest Politics: The Evolution of International Cooperation* (Routledge: London and New York).

Hund, G and O Elkhamri. 2005. "Industry Self-Regulation as a Means to Promote Nonproliferation," Report No. PNNL-15355 (Pacific Northwest National Laboratory: Seattle WA).

Hund, G and AM Seward. 2008. "Broadening Industry Governance to Include Nonproliferation," PNNL-17521 (Pacific Northwest National Laboratory: Seattle WA), cgs.pnnl.gov/pdfs/IndustryCorporateGovernance-Security_11-12-08Final.pdf.

Hvistendahl, M. 2010. "China's Hacker Army," *Foreign Policy* (Mar. 3), foreignpolicy.com/2010/03/03/chinas-hacker-army.

IASB. *See* International Association Synthetic Biology.

ICO. *See* International Coffee Organization.

ICPS. *See* International Consortium for Polynucleotide Synthesis.

IG DHS. 2008. "Code of Conduct for Nanotechnologies," www.innovationsgesellschaft.ch/media/archive2/publikationen/Factsheet_CoC_engl.pdf.

IGSC. 2012. Harmonized Screening Protocol (Amended), www.genesynthesisconsortium.org/wp-content/uploads/2012/02/IGSC-Harmonized-Screening-Protocol1.pdf.

Insight Investment. n.d. "Home Page," www.insightinvestment.com.

International Association Synthetic Biology. 2009. "Code of Conduct for Best Practices in Gene Synthesis," www.ia-sb.eu/go/synthetic-biology/activities/press-area/press-information/code-of-conduct-for-best-practices-in-gene-synthesis.

   2014. Press Release: "Code of Conduct for Best Practices in Gene Synthesis" (Feb. 3), webcache.googleusercontent.com/search?q=cache:UccsWkDSPwEJ:synthetic-biology.net/2014/02/code-of-conduct-for-best-practices-in-gene-synthesis/+&cd=1&hl=en&ct=clnk&gl=us.

International Coffee Organization. 2013. *The International Coffee Organization 1963–2013: 50 Years Serving the World Coffee Community* (ICO: London).

International Consortium for Polynucleotide Synthesis. n.d. Web Page: "About the ICPS," available from the author.

International Gene Synthesis Consortium. 2009. "Harmonized Screening Protocol," available from the author.

International Social and Environmental Accreditation and Labelling Alliance (ISEAL). 2013. "Credibility Principles Brochure: What's Essential for a Standards System to Deliver Positive Impact?" issuu.com/isealalliance/docs/credibility_principles_booklet?e=3335187/3992107.

International Social and Environmental Accreditation and Labelling Alliance (ISEAL). 2014. "Setting Social and Environmental Standards: ISEAL Code of Good Practice Version 6.0," issuu.com/isealalliance/docs/iseal_standard_setting_code_v6_dec_?e=3335187/11610938.

International Wood Markets Group. 2013. Press Release: "Canada and U.S. 'Top 20' Lumber Producers' Report Indicates Renewed Industry Optimism" (Mar. 21).

ISEAL. *See* International Social and Environmental Accreditation and Labelling Alliance.

ISIS (Institute for Science and International Security). n.d. "Case Studies of Illicit Procurement Networks," exportcontrols.info/case_studies.htm.

ITS Global. 2011. *Forest Certification – Sustainability, Governance, and Risk*, www.itsglobal.net/sites/default/files/itsglobal/Forestry%20Certification-Sustainability%20Governance%20and%20Risk%20%282011%29.pdf.

Jackson, MO. 2008. *Social and Economic Networks* (Princeton University Press: Princeton and Oxford UK).

Jaffee, D and T Russell. 2009. "Responding to WMD Terrorism Threats: The Role of Insurance Markets," pp. 251–286 in SM Maurer (ed.), *WMD Terrorism: Science and Policy Choices* (MIT Press: Cambridge MA).

Jones, A. 2011. "Law Blog: Is China Finally Getting Tough on Piracy?" *Wall Street Journal* (Apr. 20) (online edition), blogs.wsj.com/chinarealtime/2011/04/20/is-china-finally-getting-tough-on-piracy.

Journal Editors and Authors Group. 2003. "Statement on Scientific Publication and Security," *Science* 299: 1149.

Joyner, DH. 2004. "The Enhanced Proliferation Control Initiative: National Security Necessity or Unconstitutionally Vague?" *Georgia Journal of International and Comparative Law* 32: 107–123.

Kalfagianni, A. 2010. "The Effectiveness of Private Food (Retail) Governance for Sustainability," *Institute for Environmental Studies* 12: 1.

Kan, M. 2011. "Pledge Asks Chinese Hackers to Reject Cybertheft," *Computer World* (Sept. 16), www.computerworld.com/article/2511219/security0/pledge-asks-chinese-hackers-to-reject-cybertheft.html.

Katz, MA and C Shapiro. 1994. "Systems Competition and Network Effects," *Journal of Economic Perspectives* 8: 93–115.

Kelle, A. 2009. "Synthetic Biology and Biosecurity: From Low Levels of Awareness to a Comprehensive Strategy," *EMBO Reports* 10: S23–S27, www.nature.com/embor/journal/v10/n1s/pdf/embor2009119.pdf.

Kennedy, D. 2005. "Better Never Than Late," *Science* 310: 195.

King, AA and MJ Lenox. 2000. "Industry Self-Regulation Without Sanctions: The Chemical Industry's Responsible Care Program," *Academy of Management Journal* 43: 698.

King, AA, MJ Lenox, and ML Barnett. 2002. "Strategic Responses to the Reputation Commons Problem," pp. 393–406 in AJ Hoffman and MJ Ventresca (eds.), *Organizations, Policy, and the Natural Environment: Institutional and Strategic Perspectives* (Stanford University Press: Redwood Village CA).

Kirby, WT and TC Weymouth. 1985. "Antitrust and Amateur Sports: The Role of Noneconomic Values," *Indiana Law Journal* 61: 31.

Kjaer, AM. 2004. *Governance* (Polity Press: UK).

Kolk, A. 2005. "Corporate Social Responsibility in the Coffee Sector: The Dynamics of MNC Responses and Code Development," *European Management Journal* 23: 228.

— 2012. "Towards a Sustainable Coffee Market: Paradoxes Faced by a Multinational Company," *Corporate Social Responsibility and Environmental Management* 19: 79.

— 2013. "Mainstreaming Sustainable Coffee," *Sustainable Development* 21: 324.

Kosoff, Maya. 2016. "Grubhub Shares Plunge After C.E.O. Sends Memo Telling Employees Who Share Trump's Views to Quit," *Vanity Fair* November 11.

Kotchen, MJ and K van 't Veld. 2009. "An Economics Perspective on Treating Voluntary Programs as Clubs," in pp. 67–85 M Potoski and A Prakash (eds.), *Voluntary Programs: A Club Theory Perspective* (MIT Press: Cambridge MA).

Kreuter, N. 2012. "Salary Realities," *Inside Higher Ed* (Sept. 5), www.insidehighered.com/advice/2012/09/05/essay-what-new-faculty-members-need-know-about-salaries.

Krimsky, S. 1982. *Genetic Alchemy: The Social History of the Recombinant DNA Controversy* (MIT Press: London and Cambridge).

Künkel, P, V Fricke, and S Cholakova. 2008. "The Common Code for the Coffee Community (4C)," pp. 85–88 in D. Vollmer, *Enhancing the Effectiveness of Sustainability Partnerships: Summary of a Workshop*, (National Academies Press: Washington DC), www.nap.edu/openbook.php?record_id=12541&page=85.

Kydd, AH. 2015. *International Relations Theory: The Game Theoretic Approach* (Cambridge University Press: New York).

LaChappelle, J. 2013. Blog: "Landing the Big Fish: 5 Questions with MSC about the McDonald's USA Announcement" (ISEAL Alliance) (Mar. 19), www.isealalliance.org/online-community/blogs/landing-the-big-fish-5-questions-with-msc-about-the-mcdonald%E2%80%99s-usa-announcement.

Landeweerd, L, D Townend, J Mesman, and I Van Hoyweghen. 2015. "Reflections on Different Governance Styles in Regulating Science: A Contribution to 'Responsible Research and Innovation,'" *Life Sciences, Society and Policy* 11: 8.

Langley, M. 2014. "Inside Target, CEO Struggles to Regain Shoppers' Trust," *Wall Street Journal* (Feb. 19).

— 2016. "Tech CEO Turns Rabble-Rouser," *Wall Street Journal* (May 3).

Langner, R. 2013. "To Kill a Centrifuge: A Technical Analysis of What Stuxnet's Creators Tried to Achieve," www.langner.com/en/wp-content/uploads/2013/11/To-kill-a-centrifuge.pdf.

Lear, J. 1978. *Recombinant DNA: The Untold Story* (Crown: New York).

Leary, TB. 2004. FTC Chairman's Remarks: "Self-Regulation and the Interface Between Consumer Protection and Antitrust" (Jan. 28), www.ftc.gov/public-statements/2004/01/self-regulation-and-interface-between-consumer-protection-and-antitrust.

Leibowitz, J. 2005. Speech: "The Good, the Bad and the Ugly: Trade Associations and Antitrust" (Mar. 30) www.ftc.gov/public-statements/2005/03/good-bad-and-ugly-trade-associations-and-antitrust.

Leighley, JE and J Nagler. 2013. *Who Votes Now? Demographics, Issues, Inequality, and Turnout in the United States* (Princeton University Press: Princeton NJ).

Lemon Committee. *See* Committee on Advances in Technology and the Prevention of Their Application to Next Generation Biowarfare Threats.

Lenox, MJ and J Nash. 2003. "Industry Self-Regulation and Adverse Selection: A Comparison Across Four Trade Association Programs," *Business Strategy and Environment* 12: 343.

Locke, RM. 2013. *The Promise and Limits of Private Power: Promoting Labor Standards in a Global Economy* (Cambridge University Press: New York).

Lok, C. 2009. "Gene-Makers Put Forward Security Standards," *Nature News* (Nov. 4), www.nature.com/news/2009/091104/full/news.2009.1065.html.

Lu, Y. 2014. "White Knights of Cyber Security," *Global Times* (Jun. 2), www.globaltimes.cn/content/863423.shtml.

Luongo, KN and I Williams. 2007. "The Nexus of Globalization and Next-Generation Nonproliferation: Tapping the Power of Market-Based Solutions," *Nonproliferation Review* 14(3): 459–470.

Lusk, J. 2016. "Can I Get That with Extra GMO?" *Wall Street Journal* (Apr. 26).

Lynch, C. 2015. "Shutting Down Iran's Nuclear Smugglers," *Foreign Policy* (Jul. 1).

Lytton, TD. 2014. "Competitive Third-Party Regulation: How Private Certification Can Overcome Constraints that Frustrate Government Regulation," *Theoretical Inquiries in Law*, 15: 539.

Lytton, TD and LK McAllister. 2014. "Oversight in Private Food Safety Auditing: Addressing Auditor Conflict of Interest," *Wisconsin Law Review*: 289–335.

Maher, I. 2011. "Competition Law and Transnational Private Regulatory Regimes: Marking the Cartel Boundary," *Journal of Law and Society* 38(1): 119.

Maitland, I. 1985. "The Limits of Business Self-Regulation," *California Management Review* XXVII(3): 132–147.

Mann, T. 2013. "Rail Safety and the Value of a Life," *Wall Street Journal* (Jun. 17), online. wsj.com/news/articles/SB10001424127887323582904578485061024790402.

Matthews, O. 2015. "Russia's Greatest Weapon May Be Its Hackers," *Newsweek* (May 7).

Mattli, W and N Woods. 2009. "In Whose Benefit? Explaining Regulatory Change in Global Politics," pp. 1–43 in W Mattli and N Woods (eds.), *The Politics of Global Regulation* (Princeton University Press: Princeton NJ).

Maurer, SM. 2006. "Inside the Anticommons: Academic Scientists' Struggle to Build a Commercially Self-Supporting Human Mutations Database, 1999–2001," *Research Policy* 35: 839–853.

Maurer, SM. 2009. "Technologies of Evil: Chemical, Biological, Radiological, and Nuclear Weapons," pp. 47–110 in S Maurer (ed.), *WMD Terrorism: Science and Policy Choices* (MIT Press: Cambridge MA).

Maurer, SM. 2011a. "End of the Beginning or Beginning of the End? Synthetic Biology's Stalled Security Agenda and the Prospects for Restarting It," *Valparaiso University Law Review* 45(4): 1387.

2011b. *Regulation Without Government* (Nomos: Baden-Baden FRG).

2012. "The Penguin and the Cartel: Rethinking Antitrust and Innovation Policy for the Age of Commercial Open Source," *Utah Law Review* 2012(1): 269.

Maurer, SM. 2014. " Public Problems, Private Answers: Reforming Industry Self-Governance Law for the 21st Century, " *DePaul Business and Commercial Law Journal* 12(3) (Spring).

Maurer, SM and S von Engelhardt. 2013. "Industry Self-Governance: A New Way to Manage Dangerous Technologies," *Bulletin of the Atomic Scientists* (May).

Maurer, SM, R Firestone, and C Scriver. 2000. "Science's Neglected Legacy," *Nature* 405: 117.

Maurer, SM and M Fischer. 2010. "How to Control Dual-Use Technologies in the Age of Global Commerce," *Bulletin of the Atomic Scientists* (Jan./Feb.).

Maurer, SM, M Fischer, H Schwer, C Stähler, and P Stähler. 2009. "Making Commercial Biology Safer: What the Gene Synthesis Industry Has Learned About Screening Customers and Orders" (ITHS Working Paper), gspp.berkeley.edu/assets/uploads/page/Maurer_IASB_Screening.pdf.

Maurer, SM, K Lucas, and S Terrell. 2006. "From Understanding to Action: Community-Based Options for Increasing Safety and Security in Synthetic Biology" (ITHS Working Paper).

Maurer, SM and S Scotchmer. 1999. "Database Protection: Is It Broken and Should We Fix It?" *Science* 284: 1129.

2004. "Procuring Knowledge," pp. 1–31 in GD Libecap (ed.), *Intellectual Property and Entrepreneurship: Advances in the Study of Entrepreneurship, Innovation and Economic Growth*, vol. 15, Elsevier: Amsterdam.

2007."Profit Neutrality in Licensing: The Boundary Between Antitrust Law and Patent Law," *American Law and Economics Review* 8: 476.

2014. "The Essential Facilities Doctrine: The Lost Message of Terminal Railroad," *California Law Review: Circuit* (Oct. 8): 278–316.

Maurer, SM and L Zoloth. 2007. "Synthesizing Biosecurity," *Bulletin of the Atomic Scientists* 63(6).

Maxwell, JW, TP Lyon, and SC Hackett. 2000. "Self-Regulation and Social Welfare: The Political Economy of Corporate Environmentalism." *Journal of Law and Economics* 43(2): 583–617.

May, M. 2010. "Seeking Security for Synthetic Genes," *Scientific American Worldview*, www.saworldview.com/article/seeking-security-for-synthetic-genes.

May, B, D Leadbitter, and M Weber. 2003. "The Marine Stewardship Council (MSC): Background, Rationale, and Challenges," pp. 14-33 in B Phillips, T Ward, and C Chaffee (eds.), *Ecolabelling in Fisheries: What Is It All About?* (Blackwell Science: Oxford).

Mayer, F and G Gereffi. 2010. "Regulation and Economic Globalization: Prospects and Limits of Private Governance," *Business and Politics*: 12(3).

McCubbins, MD, RG Noll, and BR Weingast. 1987. "Administrative Procedures as Instruments of Political Control," *Journal of Law, Economics and Organization* 3(2): 243–277.

McCubbins, MD and T Schwartz. 1984. "Congressional Oversight Overlooked: Police Patrols Versus Fire Alarms," *American Journal of Political Science* 28: 165–179.

McElwee, S. 2014. "Why the Voting Gap Matters," *Demos* (Oct. 23), www.demos.org/publication/why-voting-gap-matters.

McGinnis, J. 1969. *The Selling of the President 1968* (Simon & Schuster: New York).

Mechel, F, N Meyer-Ohlendorf, P Sprang, and RG Tarasofsky. 2006. "Public Procurement and Forest Certification: Assessing the Implications for Policy, Law, and International Trade," *Law and International Trade*, www.ecologic.eu/sites/files/project/2013/933_final_report.pdf.

Meidinger, E. 2006. "The Administrative Law of Global Private-Public Regulation: The Case of Forestry," *European Journal of International Law* 17: 47.

2007. "Beyond Westphalia: Emerging Transnational Regulatory Systems," pp. 121–143 in C Brütsch and D Lehmkuhl (eds.), *Law and Legalization in Transnational Relations* (Routledge: Abingdon UK).

2008. "Competitive Supragovernmental Regulation: How Could It Be Democratic?" *Chicago Journal of International Law* 8(2): 513–534.

Microsoft. 2007. "New Study Finds 14.7 Million Jobs Created Globally by Microsoft and Its Ecosystem" (Oct. 18), news.microsoft.com/2007/10/18/new-study-finds-14-7-million-jobs-created-globally-by-microsoft-and-its-ecosystem/#sm.0012apazy18hrf2wraa10b3axpmdr.

Millon, D. 1991 [1988]. "The Sherman Act and the Balance of Power," pp. 85–115 in ET Sullivan (ed.), *Political Economy of the Sherman Act: The First One Hundred Years* (Oxford University Press: New York).

Milne, CP and J Tait. 2009. "Evolution Along the Government-Governance Continuum: Impacts of Regulation on Medicines Innovation in the United States," pp. 107–132 in C Lyall, T Papaioannou, and J Smith (eds.), *Limits to Governance: The Challenge for Policymaking in the Life Sciences* (Ashgate: Farnham UK).

MIT News Office. 2005. Press Release: "Study to Explore Risks, Benefits of Synthetic Genomics," web.mit.edu/newsoffice/2005/syntheticbio.html.

Moe, TM. 1998. "The Presidency and the Bureaucracy: The Presidential Advantage," pp. 437–468 in M Nelson (ed.), *The Presidency and the Political System*, 5th ed., (CQ: Washington DC).

Morris, FA, AM Seward, and AJ Kurzrok. 2012. "A Nonproliferation Third Party for Dual-Use Industries – Legal Issues for Consideration," Report No. PNNL-21908 (Pacific Northwest National Laboratories: Seattle).

MSC. *See* Marine Stewardship Council.

Nanotechnology Industries. n.d. "Home Page," www.nanotechia.org.

Nanowerk. 2008. "Swiss Retailers Introduce the World's First Code of Conduct for Nanotechnology in Consumer Products" (Apr. 18), www.nanowerk.com/news/newsid=5375.php.

n.d. "Developing a Nanotechnology Code for European Industry," www.nanowerk/com/news/newside=2841.php.

Nassauer, S. 2015. "Wal-Mart Shrinks the Big Box, Vexing Vendors," *Wall Street Journal* (Oct. 25).

National Academy of Sciences Committee on Genomic Databases for Bioterrorism. 2004. *Biotechnology Research in an Age of Terrorism* (National Research Council: Washington DC).

National Academy of Sciences Committee on Genomic Databases for Bioterrorism Threat Agents (Fink Committee). 2004. *Seeking Security: Pathogens,*

Open Access, and Genome Databases 25–27 (National Research Council: Washington DC).

National Science Advisory Board for Biosecurity. 2006. "Addressing Biosecurity," osp.od.nih.gov/sites/default/files/resources/Final_NSABB_Report_on_Synthetic_Genomics.pdf.

Nelson, R. 2015. "The Secret Republicans of Silicon Valley," *National Journal* (Apr. 8).

Newman, M. 2016. Web Page: "Starting Spots Still in the Air in NL All-Star Voting," MLB.com (Jun. 15), m.mlb.com/news/article/184177190/national-league-all-star-voting-update.

Nguyen, TH. 2005. "Microchallenges of Chemical Weapons Proliferation," *Science* 390: 1021.

Noble, RK. 2013. "Keeping Science in the Right Hands," *Foreign Affairs* (Nov./Dec.).

NTI. *See* Nanotechnology Industries.

NuclearPrinciples.org. n.d. "Nuclear Power Plant and Reactor Exporters' Principles of Conduct – Participants," nuclearprinciples.org/participants.

Ostrom, E. 1990. *Governing the Commons: The Evolution of Institutions for Collective Action* (Cambridge University Press: New York).

Ottolenghi, E. 2013. "Iran Is *Really* Good at Evading Sanctions," *The Tower*, www.thetower.org/article/iran-is-really-good-at-evading-sanctions.

Overdevest, C and J Zeitlin. 2014. "Assembling an Experimentalist Regime: Transnational Governance Interactions in the Forest Sector," *Regulation & Governance* 8: 22–48.

Oxonica Corp. 2007. "Industry Leaders to Develop Nanotech Code of Conduct" (Aug. 9), www.advfn.com/news_NanoTech-Code-of-Conduct_21800513.html.

Ozoliņa, Ž, C Mitcham, J Stilgoe, R Andanda, M Kaiser, L Nielen, N Stehr, and R Qiu. 2009. "Global Governance of Science," European Commission Report EUR 23616 EN.

Pan-European Forest Certification Council. *See* PEFC.

Parens, E, J Johnston, and J Moses. 2009. "Ethical Issues in Synthetic Biology: An Overview of the Debates" (Woodrow Wilson Center: Washington DC), www.synbioproject.org/process/assets/files/6334/synbio3.pdf.

Pearson, GR. 2006. "The Iraqi Biological Weapons Program," pp. 169–190 in M Wheelis, L Rozsa, and M Dando (eds.), *Deadly Cultures: Biological Weapons Since 1945* (Harvard: Cambridge MA).

PEFC. 2004. Web Page: "Leadership Interview with Ben Gunneberg," pefc.org/news-a-media/general-sfm-news/61-leadership-interview-with-ben-gunneberg-pefc-s-secretary-general.

n.d.(a). "Facts and Figures," www.pefc.org/about-pefc/who-we-are/facts-a-figures.

n.d.(b). Web Page: "What Makes PEFC Unique," pefc.co.uk/about-pefc/what-makes-pefc-unique.

Pennisi, E. 2005. "Synthetic Biology Remakes Small Genomes," *Science* 310: 769.

Perkovich, G and B Radzinsky. 2012. "A Common High Standard for Nuclear Power Plant Exports: Overview and Analysis of the Nuclear Power Plant Exporters' Principles of Conduct," *Nuclear Law Bulletin* 90(2): 7–22.

Peters, BG and J Pierre. 2010. "Public-Private Partnerships and the Democratic Deficit: Is Performance-Based Legitimacy the Answer?" pp. 41–54 in M Bexell and U Mörth (eds.), *Democracy and Public-Private Partnerships in Global Governance* (Palgrave: New York).

Peters, M and RE Silverman. 2016. "Big Business Speaks Up on Social Issues," *Wall Street Journal* (Apr. 17).

Pickering, A. 1984. *Constructing Quarks: A Sociological History of Particle Physics* (University of Chicago Press: Chicago).
Pitofsky, R. 1998. "FTC Chairman Address to D.C. Bar Association Symposium on Self-Regulation and Antitrust" (Feb. 19).
Pizer, WA, R Morgenstern, and J Shih. 2008. "Evaluating Voluntary Climate Programs in the United States," Resources for the Future Discussion Paper 08-13.
Polanyi, M. 1962 [1942]. "The Republic of Science: Its Political and Economic Theory," *Minerva* 1: 54.
Pollack, A. 2016. "Scientists Talk Privately About Creating a Synthetic Human Genome," *New York Times* (May 13).
Pollock, L. 2016. "Churchill on Trump and Clinton," *Wall Street Journal* (Oct. 7).
Ponsoldt, JF. 1981. "The Application of Sherman Act Antiboycott Law to Industry Self-Regulation: An Analysis Integrating Nonboycott Sherman Act Principles," *Southern California Law Rev.* 55: 1.
Ponte, S, P Gibbon, and J Vestergaard. 2011. "Governing through Standards: An Introduction," pp. 1–24 in S Ponte, P Gibbon, and J Vestergaard (eds.), *Governing through Standards: Origins, Drivers and Limitations* (Palgrave: New York).
Potts, J. 2004. "Multi-Stakeholder Collaboration for a Sustainable Coffee Sector," www.iisd.org/pdf/2004/sci_coffee_background2.pdf.
Prakash, A. 2000. "Responsible Care: An Assessment," *Business and Society*, 39(2): 183–209.
Pritchett, Jon L., and Tiryakian, E. 2017. "When CEOs Play Politics, Shareholders Can Take Them to Court," *Wall Street Journal* (Aug. 18).
Pulitzer, R and CH Grasty. 1919. "Forces at War in Peace Conclave," *New York Times* (Jan. 18).
Purbawiyatna, A and M Simula. 2008. *Developing Forest Certification: Towards Increasing the Comparability and Acceptability of Forest Certification Systems Worldwide* (International Tropical Timber Organization: London).
Purnhagen, K. 2014. "Mapping Private Regulation – Classification, Market Access and Market Closure Policy, and Law's Response," Wageningen University Law and Governance Group Working Paper 2014/04.
Putnam, GH. 1896. *Books and Their Makers During the Middle Ages* (Putnam: New York).
Rabinow, P. 2011. *The Accompaniment: Assembling the Contemporary* (University of Chicago Press: Chicago).
Rabinow, P and G Bennett. 2012. *Designing Human Practices: An Experiment with Synthetic Biology* (University of Chicago Press: Chicago).
Rappert, B. 2004. "Responsibility in the Life Sciences: Assessing the Role of Personal Codes," *Biosecurity and Bioterrorism* 2(3): 164–74.
Rawls, J. 1999 [1971]. *A Theory of Justice* (Harvard University Press: Cambridge MA).
Relman Committee. *See* Committee on Advances in Technology and the Prevention of Their Application to Next Generation Biowarfare Threats.
Responsible Nanocode. 2008. "Information on the Responsible Nano Code Initiative," www.nanocap.eu/Flex/Site/Download.aspx?ID=2736.
Responsible NanoCode. n.d. "The Working Group Participants," responsiblenanocode.org/pages/participants/index.html.
Reynolds, A. 2006. *Income and Wealth*, 2d ed. (Greenwood Press: Westport CT).

Rhodes, R. n.d. "Richard Rhodes on: Edward Teller's Role in the Oppenheimer Hearings," www.pbs.org/wgbh/amex/bomb/filmmore/reference/interview/rhodes12.html.

Richelson, JT. 2006. *Spying on the Bomb: American Nuclear Intelligence from Nazi Germany to Iran and North Korea* (Norton: New York).

Ripley, RB and GA Franklin. 1976. *Congress, the Bureaucracy, and Public Policy* (Dorsey Press: Homewood IL).

Rogers, M. 1977. *Biohazard* (Knopf: New York).

Roheim, CA, F Asche, and J Insignares Santos. 2011. "The Elusive Price Premium for Ecolabelled Products: Evidence from Seafood in the UK Market," *Journal of Agricultural Economics* 62: 655–668.

Rose, F. 2016. "Notable and Quotable: The Milton Friedman Prize," *Wall Street Journal* (Jun. 1).

Rotherham, T. 2011. "Forest Management Certification Around the World – Progress and Problems," *Forestry Chronicle* 87: 603.

Royal Society and Wellcome Trust. 2004. "Do No Harm: Reducing the Potential for the Misuse of Life Science Research," royalsociety.org/~/media/Royal_Society_Content/policy/publications/2004/9671.pdf.

Rudder, CE. 2008. "Private Governance as Public Policy: A Paradigmatic Shift," *Journal of Politics* 70(4): 899–913.

Russell, AL. 2014. *Open Standards and the Digital Age: History, Ideology, and Networks* (Cambridge University Press: Cambridge).

Russolillo, S. 2016. "GrubHub CEO Backtracks on Trump Criticism, But Damage Done," *Wall Street Journal* (Nov. 14).

RussSoft Association. 2015. "Export of Russian Software Development Industry – 12th Annual Survey," russoft.org/docs/?doc=3358.

Rutherford, GW and SM Maurer. 2009. "The New Bioweapons: Infectious and Engineered Diseases," pp. 111–138 in SM Maurer (ed.), *WMD Terrorism: Science and Policy Choices* (MIT Press: Cambridge MA).

Safdar, K. 2017. "How Target Botched Its Response to North Carolina Bathroom Law," *Wall Street Journal*. (Apr. 5 2017)

Sasser, EN. 2002. "Gaining Leverage: NGO Influence on Certification Institutions in the Forest Products Sector," pp. 229ff in LD Teeter, B Cashore, and D Zhang (eds.), *Forest Policy for Private Forestry* (CABI: New York).

Scarpa, C. 1999. "The Theory of Quality Regulation and Self-Regulation: Towards an Application to Financial Markets," pp. 236–260 in B Bortolotti and G Fiorentini (eds.), *Organized Interests and Self-Regulation* (Oxford University Press: New York and Oxford UK).

Schaller, S. 2007. The Democratic Legitimacy of Private Governance: An Analysis of the Ethical Trading Initiative, INEF Report 91/2007, Institute for Development and Peace (University Duisburg-Essen: Duisburg FRG).

Scharpf, FW. 1999. *Governing in Europe: Effective and Democratic?* (Oxford University Press: London).

Scherer, FM. 1970. *Industrial Market Structure and Economic Performance*, 1st ed. (Rand McNally: Chicago).

Scherer, FM and D Ross. 1990. *Industrial Market Structure and Economic Performance*, 3d ed. (Houghton-Mifflin: Boston).

Schleifer, P and M Bloomfield. 2015. "When Institutions Fail: Legitimacy, (De)legitimation, and the Failure of Private Governance Systems," Robert Schuman Centre for Advanced Studies Working Paper 2015/36.

Schwartzel, E and C McWhirter. 2015. "Group Backed by Koch Brothers Takes Aim at Tax Credits for Films," *Wall Street Journal* (Mar. 25).

Scotchmer, S. 2004. *Innovation and Incentives* (MIT Press: Cambridge MA).

Scotchmer, S and J Farrell. 1988. "Partnerships," *Quarterly Journal of Economics* 103(2): 279.

Scott, C, F Cafaggi, and L Senden. 2011. "The Conceptual and Constitutional Challenge of Transnational Private Regulation, *Journal of Law and Society* 38(1): 1.

Seib, JE. 1985. "Antitrust and Nonmarket Goods: The Supreme Court Fumbles Again," *Washington Law Review* 60: 721.

Selgelid, MJ. 2007. "A Tale of Two Studies: Ethics, Bioterrorism, and the Censorship of Science," *Hastings Center Reports* (May–Jun.): 35.

Service, RF. 2006. "Synthetic Biologists Debate Policing Themselves," *Science* 312: 1116.

SFI. *See* Sustainable Forestry Initiative.

Shapiro, C. 1983. "Premiums for High Quality Products as Returns to Reputations," *Quarterly Journal of Economics* 98(4): 659.

Shiell, A and D Chapman. 2000. "The Inertia of Self-Regulation: A Game-Theoretic Approach to Reducing Passive Smoking in Restaurants," *Social Science and Medicine* 51: 1111.

Short, JL. 2013. "Self-Regulation in the Regulatory Void: 'Blue Moon' or 'Bad Moon'?" *Annals of the AAPSS* 649.

Simirenko, L, M Harmon-Smith, A Visel, EM Rubin, and NJ Hillson. 2015. "The Joint Genome Institute's Synthetic Biology Internal Review Process," *Journal of Responsible Innovation* (Jan. 27).

Simpson, C. 2014. "NBA Commissioner Adam Silver Bans Donald Sterling for Life," *The Atlantic* (Apr. 29).

Sinclair, D. 1997. "Self-Regulation Versus Command and Control? Beyond False Dichotomies," *Law and Policy* 19: 529.

Singer, MD. n.d. The Maxine Singer Papers, profiles.nlm.nih.gov/ps/retrieve/Narrative/DJ/p-nid/218/p-docs/true.

Sloan, J. 2013. "Market Outlook: Surplus in Carbon Fiber's Future?" *CompositesWorld* (Mar. 1), www.compositesworld.com/articles/market-outlook-surplus-in-carbon-fibers-future.

Steering Committee of the State-of-Knowledge Assessment of Standards and Certification. 2012. *Toward Sustainability: The Roles and Limitations of Certification (Final Report)*, www.resolv.org/site-assessment/towardsustainability.

Steinbruner, JD and ED Harris. 2003. "Controlling Dangerous Pathogens," *Issues in Science and Technology* (Spring) www.issues.org/19.3/steinbruner.htm.

Stemerding, D, H de Vried, B Walhout, and R Van Est. 2009. "Synthetic Biology and the Role of Civil Society Organizations Shaping the Agenda and Arena of the Public Debate," pp. 155ff. in M. Schmidt, A Kalle, A Ganuli-Mitra, and H de Vried (eds.), *Synthetic Biology: The Technoscience and Its Societal Consequences* (Springer: New York and London).

Stoyanov, R. 2015. "Russian Financial Cybercrime: How It Works," securelist.com/files/2015/11/Kaspersky_Lab_cybercrime_underground_report_eng_v1_0.pdf.

Strassel, K. 2016. *The Intimidation Game: How the Left Is Silencing Free Speech* (Hachette: New York and Boston).

Stringham, E. 2016. *Private Governance: Creating Order in Economic and Social Life* (Oxford University Press: New York).

Suchman, MC. 1995. "Managing Legitimacy: Strategic and Institutional Approaches," *Academy of Management Review* 20(3): 571–610.

Sunstein, CR. 1999. "Winner-Take-Less Codes: The Case of Private Broadcasting," *University of Chicago Law School Roundtable* 6: 39.

Sustainable Forestry Initiative. 2009. Web Page: "International Acceptance Through PEFC," www.sfiprogram.org/files/pdf/pefc-international-2009-02pdf.

Synberc Collaboration. 2015. "Synberc: Ten Years at the Genesis of Synthetic Biology," www.synberc.org.

Thomas.net. n.d. "Bioreactor Suppliers," www.thomasnet.com/products/bioreactors-5288006-1.html.

Tomasky, M. 2010. "Turnout: Explains a Lot" (Nov. 3), www.theguardian.com/commentis free/michaeltomasky/2010/nov/03/us-midterm-elections-2010-turnout-says-a-lot.

Tozzi, J. 2009. "Watchdogs: Shell Schemes Are on the Rise." *BusinessWeek.com* (Nov. 5) www.businessweek.com/smallbiz/content/nov2009/sb20091115_791003.htm.

Tucker, JB. 2010. "Double-Edged DNA: Preventing the Misuse of Gene Synthesis," *Issues in Science and Technology* 26(3), issues.org/26-3/tucker-2.

Tumpey, TM, CF Basler, PV Aguilar, et al. 2008. "Characterization of the Reconstructed 1918 Spanish Influenza Pandemic Virus," *Science* 310: 77–80.

Unilever Corp. 2003 *Fishing for the Future II: Unilever's Fish Sustainability Initiative (FSI).*

United Nations Food and Agriculture Organisation (FAO). 2009. *Guidelines for the Ecolabelling of Fish and Fishery Products from Marine Capture Fisheries.*

U.S. Census Bureau. n.d.(a). "Voting and Registration in the Election of November 2008 – Detailed Tables," www.census.gov/hhes/www/socdemo/voting/publications/p20/2008/Table%2007.xls.

    n.d.(b). "Reported Voting and Registration of Family Members, by Age and Family Income: November 2010," www.census.gov/hhes/www/socdemo/voting/publications/p20/2010/Table7_2010.xls.

U.S. Department of Health and Human Services. 2009. "Screening Framework Guidance for Synthetic Double-Stranded DNA Providers, *Federal Register* Vol. 74 at p. 62820 (Oct. 13).

U.S. Department of Health and Human Services. 2010. "Screening Framework Guidance for Synthetic Double-Stranded DNA Providers, *Federal Register* Vol. 75 at p. 62820 (Oct. 13). Available at http://www.phe.gov/Prepredness/legal/guidance/sydna/Documents/synda-guidance.pdf.

U.S. Federal Elections Commission. 2012. "Federal Elections 2012," www.fec.gov/pubrec/fe2012/federalelections2012.pdf.

U.S. Presidential Commission for the Study of Bioethical Issues. 2010. "New Directions: The Ethics of Synthetic Biology and Emerging Technologies," www.bioethics.gov/documents/synthetic-biology/PCSBI-Synthetic-Biology-Report-12.16.10.pdf.

U.S. Federal Trade Commission and Department of Justice. 2000. "*Antitrust Guidelines for Collaborations Among Competitors*" (US Government Printing Office: Washington DC).

Vanderburgh, MP 2006/2007. "The New Wal-Mart Effect: The Role of Private Contracting in Global Governance," *UCLA Law Review* 54: 913.

van der Kloet, J and T Havinga. 2008. "Private Food Regulation from a Regulatee's Perspective," Nigmegen Sociology of Law Working Papers Series No. 2008/07.

Varian, H and C Shapiro. 1999. *Information Rules* (Harvard Business School: Boston MA).

Verbruggen, P. 2013. "Gorillas in the Closet? Public and Private Actors in the Enforcement of Transnational Private Regulation," *Regulation and Governance* 7: 512–532.

VerifiedVoting.org. n.d. "Internet Voting," www.verifiedvoting.org/resources/internet-voting.

Verkuil, PR. 2006. "Public Law Limitations on Privatization of Government Functions," *North Carolina Law Review* 84: 397.

Visseren-Hamakers, IJ and P Glasbergen. 2006. "Partnerships in Forest Governance," *Global Environmental Change* 17: 408.

Vogel, D. 2005. *The Market for Virtue: The Potential and Limits of Corporate Social Responsibility* (Brookings Institution Press: Washington DC).

Vogel, D. 2010. "Taming Globalization? Civil Regulation and Corporate Capitalism," pp. 472–494 in D Coen, W Grant, and G Wilson (eds.), *The Oxford Handbook of Business and Government* (Oxford University Press: New York).

von Engelhardt, S and SM Maurer. 2012. "Industry Self-Governance and National Security: On the Private Control of Dual-Use Technologies," Goldman School of Public Policy Working Paper GSPP12-005.

Wade, N. 1975. "Genetics: Conference Sets Strict Controls to Replace Moratorium," *Science* 187: 931–935.

1977. *The Ultimate Experiment: Man-Made Evolution* (Walker & Co.: New York).

Wadman, M. 2009. "US Drafts Guidelines to Screen Genes," *Nature News* (Dec. 4).

Waldman, M. 2016. *The Fight to Vote* (Simon & Schuster: New York).

Ward, TJ. 2008. "Measuring the Success of Seafood Ecolabelling," pp. 207–243 in T Ward and B Philips (eds.), *Seafood Ecolabelling: Principles and Practice* (Wiley: Hoboken NJ).

Weart, SR. 1976. "Scientists with a Secret," *Physics Today* 29(2): 29–30.

Weaver, C. 2012. "Panel Backs Publishing Bird-Flu Research," *Wall Street Journal* (Mar. 30).

Webb, K. 2004. "Voluntary Codes: Where To From Here?" in K Webb (ed.) *Voluntary Codes: Private Governance, the Public Interest and Innovation* (Carleton Research Unit for Innovation, Science and Environment).

Weber, M. 1919. "Politics as Vocation," anthropos-lab.net/wp/wp-content/uploads/2011/12/Weber-Politics-as-a-Vocation.pdf.

Wellik, R. 2012. "Global Food Safety Initiative Improves Organizational Culture, Efficiency in Food Industry," *Food Quality & Safety* (Apr./May).

Wikipedia. n.d.(a). "Charles Robert Scriver," en.wikipedia.org/wiki/Charles_Scriver.

n.d.(b). "Dominique Lorentz," en.wikipedia.org/wiki/Dominique_Lorentz.

n.d.(c). "Georges Besse," en.wikipedia.org/wiki/Georges_Besse.

n.d.(d). "Herfindahl Index," en.wikipedia.org/wiki/Herfindahl_index.

n.d.(e). "National Institutes of Health," en.wikipedia.org/wiki/National_Institutes_of_Health.
n.d.(f). "Nuclear Program of Iran," en.wikipedia.org/wiki/Nuclear_program_of_Iran.
n.d.(g). "Research Reactor," en.wikipedia.org/wiki/Research_reactor.
n.d.(h). "Richard Cotton (Geneticist)," en.wikipedia.org/wiki/Richard_Cotton_(geneticist)#The_Human_Variome_Project_.28Founder.29.
n.d.(i). "Profit (Economics)," en.wikipedia.org/wiki/Profit_(economics).
n.d.(j). "Serials Crisis," en.wikipedia.org/wiki/Serials_crisis.
n.d.(k). "Software Industry in China," en.wikipedia.org/wiki/Software_industry_in_China.
n.d.(l). "Walter Eucken," en.wikipedia.org/wiki/Walter_Eucken.
Williamson, O. 1975. *Markets and Hierarchies: Analysis and Antitrust Implications: A Study in the Economics of Internal Organization* (Free Press: New York and London).
Wilson, JQ. 1989. *Bureaucracy: What Government Agencies Do and Why They Do It* (Basic Books: New York).
Wirtz, R. 2010. "Role and Responsibility of the Civil Sector in Managing Trade in Specialized Materials," in C Hinderstein (ed.), *Cultivating Confidence: Verification, Monitoring, and Enforcement for a World Free of Nuclear Weapons* (Nuclear Threat Initiative: Washington DC).
Wisconsin Project on Nuclear Arms Control. n.d.(a). Webpage: "About Us," www.wisconsinproject.org/aboutus.html.
  n.d.(b) Webpage: "Iranwatch: Alphabetical List of Iranian Entities," www.iranwatch.org/iranian-entities.
Wolfinger, RE and SJ Rosenstone. 1980. *Who Votes?* (Yale University Press: New Haven CT).
Wolsey, M. 2007. "Interview with Robert Reich: Supercapitalism: Transforming Business," www.forbes.com/2007/09/06/book-qanda-reich-oped-cx_mw_0906reichqanda.html.
Woodrow Wilson Center Synthetic Biology Project. n.d. "Maps Inventory," www.synbioproject.org/inventories/maps-inventory.
World Nuclear Association. 2016. Web Page: "Research Reactors," www.world-nuclear.org/information-library/non-power-nuclear-applications/radioisotopes-research/research-reactors.aspx.
Wortman, D. 2002. "Shop and Save," *Sierra Magazine* (Nov./Dec.).
Zerodium Corporation. n.d. "Our Exploit Acquisition Program," www.zerodium.com/program.html.
Zilinskas, RA and JB Tucker. 2002. "Limiting the Contribution of the Open Scientific Literature to the Biological Weapons Threat," *Journal of Homeland Security* (Dec. 3), www.homelandsecurity.org/newjournal/articles/tucker.html.

Index

A. Q. Khan Network, 30, 204, 206
   and government regulators, 29
Agnew, Paul Gough, 13, 148
Anderson, Ephraim, 103, 105, 107, 108
antitrust. *See* competitions policy
Asilomar Conference, 16, 99, 115
   Asilomar I (preliminary conference), 103
   Asilomar II (main conference), 104
   moratorium, 102
Astra-Zeneca Pharmaceutical Corp. (anchor firm), x, 41, 44
ATG:biosynthetics (DNA manufacturer), 44
atomic bomb. *See* nuclear weapons
avian influenza ("bird flu"), 117, 121

Baltimore, David, 102, 103, 105
BASF (anchor firm), 36
Berg, Paul, 99, 101–8
Berners-Lee, Tim, ix, 14
biological weapons. *See* weapons of mass destruction
Biomax Informatics AG (DNA manufacturer), 44
Blue Heron Biotechnology, Inc. (DNA manufacturer), 38, 44
Botstein, David, 107, 134
Boyer, Herbert, 101
Brenner, Sidney, 104, 105, 107
Brooks, Lisa (NIH Official), 111, 114
Bumble Bee Tuna (anchor firm), 24

chemical weapons. *See* weapons of mass destruction
Chicken of the Sea (anchor firm), 24
China, 209, 210, 216

coffee initiatives
   anchor firms' market power, 30, 31
   Common Code for the Coffee Community (4C), 29
   and competition for market share, 30
   and consumer demand for sustainable products, 29, 30
   effectiveness of, 31
   government efforts to promote private coffee initiatives, 29, 30
   government intervention in coffee markets, 29
   industry economics and "preferred supplier" agreements, 30
   politics of, 30
   and staff, 29, 30, 92
Cohen, Herbert, 101, 107, 108
Collins, Francis, 113
competitions policy (antitrust)
   as deterrent to self-governance, 55, 165
   economic efficiency interpretations of, 168, 172
   essential facilities doctrine, 176
   European, 167
   Japanese, 167
   political interpretations of, 167, 168
   sham agreements, 175, 177
   United States (Sherman Act), 167
   using private standards to injure competitors, 59, 65
Constitutional and Common Law, 166
Cotton, Richard G. H. ("Dick"), ix, 110, 111, 183
Craic Computing (screening software provider), 44
Curtiss, Roy, 103, 104
cybersecurity, 215

Dixon, Robert, 13
DNA – artificial or synthetic. *See* synthetic biology initiatives – commercial
Dolphin-Safe Tuna Initiative, 23
Douglas, William O., 170
Downs, Anthony, 4, 72, 78, 80, 82, 140, 154
DuPont Chemical Co. (anchor firm), 36

Earth Island Institute (NGO), 23, 24
Endy, Drew, 116
Entelechon (DNA manufacturer), 44
Environmental Defense Fund (NGO), 36
European Bioinformatics Institute, 110
European Nanotechnology Trade Association (trade organization), 37

Falkow, Stanley, 103, 104
Farrell, Joseph, 91
Fashion Originator's Guild of America, 12–13, 169
Febit Synbio (DNA manufacturer), 44
Federal Bureau of Investigation (FBI), 42, 43
    efforts to promote private governance, xi, 42–3, 163
Fermi, Enrico, 16, 18
Fischer, Markus, 183
fisheries initiatives, 26
    and competition for market share, 27, 28
    and consumer demand for sustainable products, 27
    effectiveness of, 27, 29
    Marine Stewardship Council (MSC), 26
    politics of, 28
FOGA. *See* Fashion Originator's Guild of America
food safety initiatives, 24
    Association of UK Retailers Standard, 24
    effectiveness of, 26
    Global Food Safety Initiative (GFSI), 25
    Global G.A.P. (formerly Eurep G.A.P.), 25
    and government regulators, 26
    International Food Standard, 25
    Safe Quality Food Standard, 25
forestry initiatives, 32, 92
    competition between standards, 34
    and consumer demand for sustainable products, 32
    effectiveness of, 32, 33, 35
    Forest Stewardship Council (FSC), 32, 34
    government intervention and interactions, 32, 35
    institutional architecture and politics, 34
    Program for Endorsement of Forest Certification, formerly Program for European Forest Certification (PEFC), 33, 34
    Sustainable Forests Initiative (SFI), 33
future self-governance initiatives
    civilian, 202

Geneart Corporation, 38, 42, 44
Generay Biotech (DNA manufacturer), 44
Genome Database, 110
GenScript (DNA manufacturer), 43
Georgia Research Alliance, 113
government intervention in private initiatives
    in academic science, 191, 217
    in commercial science, 178
    cultural and organizational impediments, 194, 222
    feasibility and desirability, 182, 196, 220, 221
    in private standards wars, 190
    legitimacy, 196, 220
    managing interaction with public regulations, 188
    potential benefits, 178
    potential costs, 181
    strategy and tactics, 184, 217

HGBASE, 110
Hofstadter, Douglas, 168
Home Depot (anchor firm), 32, 181
Human Gene Mutation Database, 110
Hund, Gretchen, 205

Incyte Pharmaceutical Corp. (anchor firm), 111, 113
Insight Investment (NGO), 37
Integrated DNA Technologies, Inc. ("IDT") (DNA manufacturer), 38, 43
International Association – Synthetic Biology (IASB). *See* synthetic biology initiatives – commercial
International Consortium for Polynucleotide Synthesis (ICPS). *See* synthetic biology initiatives – commercial
International Gene Synthesis Consortium (IGSC). *See* synthetic biology initiatives – commercial
Iran, 208–10

Joliot, Frédéric, 16, 17

Keasling, Jay, 115
Kennedy, Donald, 118
Kourilsky, Philippe, 108
Kraft Foods (anchor firm), 29

Lederberg, Leon, 103, 106–8
legitimacy, 147, 148
   and academic science, 162
   benchmarks, 196
   cognitive, 161
   compared to existing governments, 155
   and "consensus" procedures, 153, 161
   corporatist, 159, 162
   and delegation, 154, 162
   deliberative, 157, 163
   demos, concept of, 149
   derivative, 161, 163
   directions for future research, 196, 222
   non-democratic, 162
   output, 160
   representative, 151, 163
   semi-empirical theories, 161
   and shadow electorates, 152, 156
   shadow of hierarchy, 13, 162
   and transparency, 154, 159, 161
   veil of ignorance, 150
Lewis, Andrew, 100, 105
Leybold GmbH, 204–7
Lowe's Home Improvement (anchor firm), 32
Lytton, Timothy, 156

Maitland, Ian, 48
March of Dimes, 109, 110, 140
markets, choice, and collective action, 2, 48
   choice between individual vs. industry-wide rules; multihoming, 50, 59, 64
   coercion and dissent, 3, 54
   imperfect markets and coercion, 52–4
   perfect markets, consumer choice, and "feel good" equilibria, 52
MDI. *See* Mutations Database Initiative
Meidinger, Errol, 35, 74, 91, 147
Mitomap, 110
MSC. *See* fisheries initiatives - Marine Stewardship Council
Mulder, Carel, 100
Mutations Database Initiative (MDI), ix, 109

nanotechnology initiatives, 36
   American, 36
   British, 37
   European, 37

National Academy of Sciences (NAS), 18, 39, 101, 102, 108, 115, 117
National Institutes of Health (NIH), 99, 100, 102, 104, 108, 111, 113, 114, 118, 120–2, 136, 140, 163
Nestlé (anchor firm), 29
Network Effects and Philosopher King Models, 14–15, 89, 99, 115
NIH. *See* National Institutes of Health
nuclear non-proliferation initiatives, 16, 203
   atomic physicists' conspiracy (1940-41), 16
   and carbon fiber, 208
   and centrifuges, 203
   and plutonium, 209
   and reactors, 209
Nuclear Power Plant Exporters group, 210
Nuclear Weapons. *See* Weapons of Mass Destruction - Nuclear

Ostrom, Elinor, xii, 127, 128, 139, 140

PEFC. *See* forestry initiatives
Pharmacia Corp., 113
Philosopher Kings. *See* Network Effects and Philosopher King Models
Pollack, Robert, 99, 103
PolyQuant (DNA manufacturer), 44
private politics – academic communities
   external audiences – congress, 140
   external audiences – general public, 124
   external audiences – regulators, 140
   players and preferences, 124, 125
   proposed reforms – ethical, legal and social implications ("ELSI") reviews, 141
   proposed reforms – NGO participation, 142
   proposed reforms – public participation, 142
   strategy and tactics – ambiguity, 124, 125, 134, 139
   strategy and tactics – controlling agendas, 133
   strategy and tactics – disinformation, 139
   theory of, 138
private politics – commercial communities
   and coalitions, 78, 83
   and "consensus", 73
   delegation and power-sharing, 50, 52, 58, 78, 79
   external politics, 59, 74, 77, 83, 87, 92
   institutions, 74, 76, 85, 86
   and limited information, 72, 76, 82
   in monopolies and cartels, 75, 80
   in oligopolies, 80
   players, preferences, and tactics, 66, 73, 75, 81

private politics – commercial communities (*cont.*)
  transparency, 74
  walk-outs, 76, 79, 83
private power. *See also* markets, choice, and collective action; shadow electorates
private power – academic communities
  and biotech firms, 135, 137, 141, 143
  enforcement by community, 135
  enforcement by governments, 135
  enforcement by journal editors, 136
  feasibility of, 123, 136
  impact of tenure and anonymity protections, 134, 136
  and merit networks, 131
  and old boy networks, 122, 131
  peer review, rewards, and "The Republic of Science", 120–2
  and shadow of hierarchy, 121
  "splitting the community" – in practice, 119
  "splitting the community" – in theory, 123, 129
  trust and reputation networks, 6, 119, 126, 128–9
private power – commercial communities
  effectiveness and impact, 50, 94
  enforcement, 15, 68
  and information asymmetry, 51
  and intelligent actor risk, 60
  interaction with government regulators and standards, 69, 75, 95, 188
  limits on. *See* shadow electorates
  power in individual supply chains, 49
  power across multiple supply chains, 52
  power and shadow of hierarchy, 13
  purchasing power and fixed costs, 51, 55
  repeat transactions and preferred supplier agreements, 51, 60, 61
private power – constraints. *See* legitimacy
private power – regulatory and litigation constraints
  government regulation, 62
  litigation, 62
  public and regulatory backlash, 63
private power – shadow electorates
  composition and preferences of, 67, 84, 85
  economics of, 50, 51, 66, 68, 84
  employees, 50, 67
  transparency and ability to make choices, 67, 73, 85

Program for Endorsement of Forest Certification, formerly Program for European Forest Certification ("PEFC"). *See* forestry initiatives

reputation networks. *See* trust and reputation networks
Royal Society, 115
Russell, Andrew, 13, 148
Russia, 210, 216

Saloner, Garth, 91
Sara Lee (anchor firm), 29, 31
Scarpa, Carlo, 15
Scriver, Charles, 110
self-governance – network effects
  and "philosopher-king" models, 14, 86, 115
self-governance – new self-governance models. *See* Dolphin-Safe Tuna; food safety; fisheries; coffee; forestry; nanotechnology; and synthetic biology initiatives
  defining characteristics, 1, 3, 23, 49, 51
self-governance – procedures, 13
  consensus, 13, 26
  transparency and due process, 13
self-governance – traditional models
  medieval guilds, 11
  standards and interoperability, 11
  shadow of hierarchy, 12
  supply chain governance, 12
Selgelid, Michael, 118
SFI. *See* forestry initiatives
shadow electorates. *See* private power – shadow electorates
Shapiro, Carl, 15
Sherman Act. *See* competitions policy
Sherman, John, 168
Shinegene Molecular Biotech (DNA manufacturer), 44
Singer, Maxine, 100–2, 106, 183
Sloning BioTechnology (DNA manufacturer), 44
social networks. *See* "trust and reputation networks."
Star-Kist (anchor firm), 23
Sustainable Forests Initiative (SFI). *See* forestry initiatives
Synberg (academic biology collaboration), 121
Synbia, 44. *See* Synthetic Biology Industry Association

synthetic biology initiatives – academic, x
Synthetic Biology 1.0 (conference), 115
Synthetic Biology 2.0 (conference), 115
synthetic biology initiatives –
  commercial, 38
  advantages of industry-wide
    standards, 39
  anchor firm pressure to adopt
    screening, 39
  available screening technologies, 40
  DNA2.0, 38, 42
  effectiveness of, 46
  industry economics and "preferred supplier"
    agreements, 40
  interaction with U.S. government regulators,
    38, 42, 45, 47
  International Association – Synthetic
    Biology (IASB), x, 41, 43
  International Consortium For
    Polynucleotide Synthesis (ICPS), 41, 42
  International Gene Synthesis Consortium
    ("IGSC"), x, 43, 44
  screening by individual suppliers, 39
Szilard, Leo, 16, 183

Teller, Edward, 135
trust and reputation networks, 6, 119, 126–8
Tschibo (anchor firm), 29
Tucker, Jonathan, 43
tuna. *See* Dolphin-Safe Tuna Initiative

UL. *See* Underwriters Laboratories
Underwriters Laboratories ("UL"), 11
Unilever (anchor firm), 26, 28

W3C, 14, 87
Walmart (anchor firm), 23, 28, 29, 181
Watson, James, 100, 102, 104, 106,
  108, 140, 141
weapons of mass destruction ("WMD")
  biological, 212–14
  chemical, 211–12
  cyber, 215–17
  nuclear weapons, 4, 16–19, 184, 203–11
  radiological, 214–15
Williamson, Oliver, 4, 72, 81, 84
Wimmer, Eckhard, 115
World Wildlife Fund (NGO), 26, 32
WWF. *See* World Wildlife Fund (NGO)

For EU product safety concerns, contact us at Calle de José Abascal, 56–1°, 28003 Madrid, Spain or eugpsr@cambridge.org.

www.ingramcontent.com/pod-product-compliance
Lightning Source LLC
LaVergne TN
LVHW041621060526
838200LV00040B/1376